T0262903

Understanding Interaction
The Relationships
Between People,
Technology, Culture, and
the Environment

Understanding Interaction
The Relationships
Between People,
Technology, Culture, and
the Environment

*Volume 1: Evolution, Technology,
Language and Culture*

Where Do We Come From and
How Did We Get Here?

Bert Bongers

CRC Press
Taylor & Francis Group
Boca Raton London New York

CRC Press is an imprint of the
Taylor & Francis Group, an **informa** business
AN AUERBACH BOOK

Cover Image Credit: Still of interactive video pattern created with a range of motion sensors in the *Facets* kaleidoscopic algorithm (based underwater footage of seaweed movement) by the author on 4 February 2010, for a lecture at Hyperbody at the Faculty of Architecture, TU Delft, NL.

First Edition published 2022
by CRC Press
6000 Broken Sound Parkway NW, Suite 300, Boca Raton, FL 33487-2742

and by CRC Press
2 Park Square, Milton Park, Abingdon, Oxon, OX14 4RN

© 2022 Taylor & Francis Group, LLC

CRC Press is an imprint of Taylor & Francis Group, LLC

Library of Congress Cataloging-in-Publication Data
A catalog record has been requested for this book.

ISBN: 978-1-4822-2862-5 (hbk)
ISBN: 978-1-032-15765-8 (pbk)
ISBN: 978-1-315-37338-6 (ebk)

DOI: 10.1201/9781315373386

Typeset in Sabon
by Deanta Global Publishing Services, Chennai, India

Contents

Preface

Technology is increasingly out of control ... and this is very exciting. The possibilities and powers of technology are increasing almost exponentially, particularly in the last hundred years. The impact of technology on people, societies, and the environment has been profound. But not always in a positive way. If we have more advanced control and rich interaction with technology, we will be able to get more work done, play more, be more creative and imaginative, express ourselves better, and learn more. In symbiotic relationships between people, technology, and the environment, humanity and indeed the whole ecosystem can reach a higher level, making the world a better place.

In order to keep (or bring) technology under our control, we need to be able to interact with it. This book is about understanding how to do that; how to invent, design, and develop appropriate technologies, that work with us and for us, in symbiotic and constructive collaborations. *Understanding Interaction* is about the relationships between people and technology, from the most basic interactions between a person and a tool, to the larger complex ecology of people, society, culture, technology, and the natural environment.

There are tensions, as expressed in the view of humans *versus* nature, but I think it is good to approach humanity as part of nature, as part of the environment (the view of many traditional cultures). We are animals in a complex ecosystem that involves nature *and* technology.

Understanding Interaction is a book that brings together perspectives and knowledge from a wide range of disciplines, synthesises this into a coherent vision, and presents frameworks applicable for design and development. It defines and explains what actually happens in the interaction between people and technology, as a basis for design and implementation of better interfaces. An important aim of the book is to demystify technology, by placing technology in a historical context and relating social and cultural developments to technical advances. It presents (new) frameworks and design tools and techniques, from theory to practice, and back again. The purpose of this book is to be a guide and a resource for designers, artists, engineers, psychologists, and social scientists. It should be useful for both novices and more experienced practitioners, for finding starting points as well as opportunities to extend knowledge and insights, drawn from different disciplines and fields.

The title of this book is a respectful reference to inspirational books in related fields such as *Understanding Media* by Marshall McLuhan in the mid-1960s,[1] and *Understanding Design* by Kees Dorst in the early 2000s.[2] The purpose of this book is to support the fostering of a deep understanding of all aspects of the interaction between people, technology, culture, and the environment.

INTERACTION AND INTERFACES

Interaction is defined as the reciprocal relationship between entities. It involves actions and reactions, the exchange of information, intentions, expressions and manipulations, effects, and potentially mutual change in behaviours and states.

Interaction, as the two-way process of input and output between people and technology, is facilitated by **interfaces**. The interface is where people meet the technology. Adequate and appropriate interfaces allow for profound ways of expression, and development of human thinking, ideas, and behaviours, and offer potential for deep interaction. A successful interface is not a superficial thing; it can't be just on the surface, added to the technology at the last stage of development. The interface has to be developed as part of the technology, from the ground up. ("The beauty of the interface is more than skin deep" as I have written earlier[3]). To allow deep interaction, the interface has to be rooted at the lowest levels of the technology, and integrated in all aspects. Designing appropriate interfaces therefore requires sufficient knowledge and understanding of both the natural and the technological environment, not just matching the technology with people but actually creating opportunities for anthropomorphic resonances.[4]

A good design(er) facilitates rather than limits, proposes rather than imposes, guides rather than fixes, suggests rather than implies, allowing for serendipity and opportunities to emerge. Understanding interaction requires insight and knowledge of what people do, their (our) behaviours, what their intentions and desires are in work, play, and explorations, what their abilities are, their social dynamics, and their relationships with the environment. It is a process of absorption, translation, adaptation, respectful engagement, and empathy.

Interaction loops, and interfaces

In this section a typical interaction loop is described, as shown in Figure I.1. In the interaction between a person and a technical device, there is usually

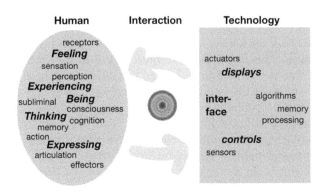

Figure I.I Diagram showing the elements of the interaction loop between a person and a system.

a form of **perception and experience** by the person, the picking up of information in the environment through their **senses** (through receptors in the sensory systems such as visual, auditory, tactual), it involves **processes of thought** (processing of information, cognition and memory, unconscious thinking, intuition, rational reasoning, formulating of intentions, behaviours), potentially externalised by the person in **expressions** through any of their output modalities (speech, movement, etc.), and in any form (modes such as language, singing, sign language, manipulation), this will be picked up by the device through its **controls and inputs** (consisting of sensors), **processed** inside the device (which can have its own processing, intelligence, and memory), and **displayed** through its outputs (actuators such as lights, screens, speakers, motors, etc.). This device output can then be picked up by the person's senses and the loop continues. The interaction can start at any point (or multiple points simultaneously), usually it is the person expressing an intention (for instance, I want the light to come on so I press a button or flick a switch, or I want to know what the weather might be tomorrow so I open up an app on my smartphone by tapping on the touchscreen). But it can also be the device that initiates the process, for instance, an alarm that goes off or a notification that is displayed (about another 'essential' software upgrade that is going to interrupt my work). The part of the technology that facilitates this interaction, making the input and output possible, is the interface. Interfaces in technical terms consist of sensors (facilitating input, control) and actuators (enabling output, display).

In this example the interaction is limited to one person and one system, but of course in reality there can be many more participants involved, multiple humans and as well as technologies, all interacting with each other – or through each other: an important function of technology is enabling people to communicate with each other, as a medium. The locus of the interaction shifts from the technology (dialling a number, typing in an email address, selecting a URL) to the interaction between people (having a phone conversation, an email exchange, looking at a website with information) in a transparent way. This is discussed in more detail in the Interface section at the end of this Volume, and in Chapter 12 *Understanding* in Volume 2.

In order to optimise the interaction loops, we need to design the technology so that it can optimally resonate with people and the environment. To do this properly a lot of knowledge is needed, not only knowledge of the technology (engineering and scientific knowledge) but also knowledge about humans, and nature. This is further discussed below, and as mentioned before this is the purpose of the whole book, to deeply understand this interaction.

The interaction between people and technology has been studied in a large range of disciplines, traditionally in Ergonomics or Human Factors, and later specifically for digital technologies in Human–Computer Interaction, Interaction Design, and more recently further specialisations (in context or approach) of Participatory Design, Experience Design, Service Design, and Social Design. All these are described and discussed in Chapter 1 *Interacting*, but at this point it is important to emphasise the purpose of the approach of *Understanding Interaction* – to study the broader picture of relationships between people and technology – not just between a person and a tool, or people and computers, but going beyond Human–Computer Interaction, increasingly engaging with the wider application of technology and society. And I am further extending this scope to the whole relationship between the human-made and natural environments.

This broader view is needed, particularly if we want to project ahead with a future-oriented approach, allowing for the emergence of innovations and novel paradigms. This book is about this broad approach, presenting new knowledge as a result of combinations of studies of traditional scientific disciplines and first principles. With this, a new picture is built up, about the relationship and interactions between people and technological artefacts, and more broadly, between the human-made technical environment and the natural environment. The aim is to break down any existing opposition between these environments, the perceived opposition between technology and nature, and by exploring synergies that deliver a hybrid approach of symbiosis.

Understanding Interaction draws on a large range of disciplines and fields of study, primarily from science, technology, and engineering (on the right-hand side of the diagram in Figure I.2), human and natural sciences (on the left-hand side of the diagram) such as sociology, psychology, physiology, biology, etc.). But it also involves design and other disciplines which traditionally always have been engaged with the relationships between people and technology, by designing and developing technology in such a way that it is useful for people, appropriate, efficient, effective, pleasurable, and flexible (in the top-middle of the diagram).

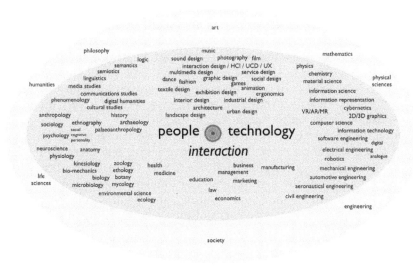

Figure I.2 Overview of the disciplines that support Understanding Interaction.

Technology in history

In everyday language it is common when using the term 'technology' to refer to digital technology (computers, smartphones, etc.). In this book the term 'technology' is defined to cover all human-made artefacts and environments, not just the latest digital technology manifested in smartphones and computers. Everything human made, every artefact, is 'technology' which we interact with – the door handle is the interface to the door, the steering wheel controls the car, the axe is held by the handle to chop wood, the chair I sit on, the guitar I play, the pencil I write with, the smartphone I fiddle with ... and yes, the computer keyboard I type this sentence on, to input my thoughts into digital data (or printed in a book). In order to understand interaction with digital technologies, particularly beyond the current paradigms of developing innovative new ways of interacting, it is helpful, essential even, to look at interaction with older, traditional technologies. Because a substantial part of the interaction through interfaces with digital devices is by physical means, its character is often reminiscent of earlier technologies. This is not always optimal, we all know the story of the mechanical typewriter which is said to have the keys arranged to actually slow down the typing (in order to avoid the mechanism jamming), and that strange coincidence of all the letters for the word 'typewriter' arranged in a neat rhythm on the upper row. Obviously there is no reason to assume this keyboard layout would be the most appropriate interface for inputting text in the computer. More optimal and efficient keyboard layouts have been developed, such as the Dvorak layout (though it only yields an increase in typing speed of 10%), and chording keyboards (but they are harder to learn as they work with combinations of a limited set of keys pressed in order to select any character).[5] Speech recognition can be used for text input, with increasing accuracy over time as the technology improves, and for giving commands, but this is not always appropriate in any social context. And personally, I still have great reservations in talking to my computer, perhaps due to my somewhat introverted nature, anyway writing is very different from speaking. More on all this later.

Before the computer keyboard there were electric typewriters, and before that mechanical ones (I learned to type on one of those in the 1970s, very hard work!). Before that, there was the printing press with movable type (one of the first examples of industrialisation) in the 15th century, before that etching and printing, and handwriting on paper, before that papyrus and parchment, and the earliest writing in stone and clay from about five thousand years ago.[6] Writing in itself was a development of drawing and painting, a development from mimetic (images resembling the things and concepts represented) to abstract (symbolic representation which has an arbitrary relationship with things and concepts). Rewind further, and all language was spoken; physical and mechanical tools were mostly used for carrying out other tasks (including making music – a language too). The first such tools were used by our ancestors millions of years ago. This is where the story starts, when our ancestors started to distinguish themselves from other apes by walking upright, which freed the hands from their previous occupation of tree climbing and (knuckle) walking. The hands gradually evolved into more dexterous extensions, capable of carrying out increasingly sophisticated manipulations of objects.[7] Walking upright also led to further anatomical changes in the larynx, enabling the first speech-like utterances.[8] With the hands exploring using tools came the notion of making and designing, and the emergence of language was driven by the need for cooperating in larger social groups. The first Volume of the book is about these developments, including an exploration of how we communicate and construct meaning. In order to understand the present, and to be able to project forward and drive future inventions, it is important to understand the past. Everything that makes our species unique is a result of a complex interrelationship and co-evolution over millions of years, involving our physical and mental capabilities, and driven by toolmaking, language, and culture, particularly in the last few hundreds of thousands of years. Agriculture, cities, writing, and sedentary civilisations had a further influence on our being, followed by industrialisation which was another major change, and finally the information age.[9]

Jumping back to the present, but with the awareness of how we arrived where we are now after many developments, it is possible to reflect on the current situation (and therefore the future) in a broader sense.

Understanding Interaction not only involves understanding techno-logical developments and the evolutionary perspective; it is also crucial to understand people. The second Volume of the book is about that, look-ing at people physically and mentally: how we perceive the outside world, how we think (or think we think), and how we interact with the environ-ment, through our senses and our actions, and with other people in social interactions.

The final chapter draws all this knowledge together, and presents (or repeats) design frameworks, approaches, and structures that are used for understanding and expanding interactions.

CONNECTING DISCIPLINES

When travelling through the many disciplines (as seen in the diagram in Figure I.2), as I have done for most of my life and particularly while writing this book, it is inevitable to feel like a dilettante at times (or even perpet-ually). Particularly in fields further removed from my own mix of back-grounds, it is often truly intimidating! In some cases having advocates and mentors from these disciplines really helped and I am grateful for the sup-port I have received, particularly Theo van Leeuwen, semiotics professor at UTS, Kees Dorst, design thinking professor at UTS, Gerrit van der Veer, HCI professor in the Netherlands, Berry Eggen, formerly at Philips and since 2003 professor at the TUE, and Bill Verplank as interaction design pioneer. All of them also read the initial book proposal (and in some cases later parts of the book), and their feedback and criticisms really helped to develop the book in its current form, which required many more radical explorations than anticipated.

One of Bill's many drawings on the initial book proposal text showed a web (as shown in Figure I.3) as a metaphor for the wider scope he encour-aged me to develop, "multiple entry points [imagine a multi-web site]". Then he extended the metaphor by rhetorically asking me "are you the spi-der or the fly" – "or the maker of the house or tree they hang on". I realised at that point, with the original book outline, I was the fly!

Many others have helped me, in projects, jobs, and even by just hav-ing conversations – my colleagues and students (many are mentioned in Chapter 1 *Interacting*), particularly at UTS in the School of Design, the Transdisciplinary School, and colleagues in architecture, built environ-ment, IT, engineering, health, and social sciences.

The original book outline unintentionally followed the journey from the foothills, the metaphor I used for my PhD book, without getting much fur-ther. It turned into something more ambitious, and with the final work aiming to reach more of the mountain range – though of course with the advantage of exploration comes the outlook on further goals to reach. But

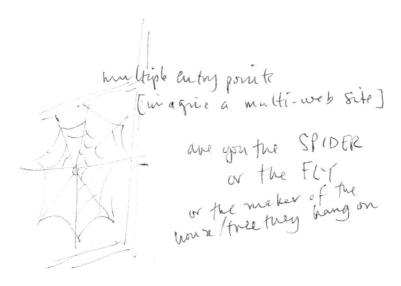

Figure I.3 Bill Verplank's drawing of the 'Understanding Interaction web'.

it was time to stop. It has taken long enough as it is, and while the original book plan could have been finished in a year or so as intended (as it was mostly based on knowledge and insights I had already gathered), it was much more fun to explore new fields even though that meant a bit more patience. It really needed more than five years to get here, after I started writing during a sabbatical semester in 2015.

By now, the sheer amount of writing reflects the complexity of an in-depth exploration of the relationship between people and technology, or between the artificial (human-made) and the natural environment. To preserve the interconnectedness of all the disciplines and topics involved in this exploration (see Bill Verplank's note in Figure I.3), I had to resist the temptation to split up the work in smaller bits, to avoid losing some of the connections. There are many topics that appear in multiple chapters or sections throughout the book, these became truly interdisciplinary bridges, as well as many cross references to connect parts of the web. They will help to navigate the maze I created.

STRUCTURE OF THE BOOK

With almost the entire book written (see overview below), it became clear that there was a natural divide between the historical part and the human science part (and the design chapter). This led to the creation of Volume 1 and Volume 2, with extensive cross-referencing between the sections and chapters of both volumes.

I have avoided the use of footnotes, many academic books have a strong undercurrent of details and distractions, in some cases the footnotes are so substantial that they form a separate (version of the) book. So I only use footnotes for the really irrelevant bits, trivia, and garden paths, while I use endnotes for the references (including the often excessive detail about these sources which might be relevant) because if one feels the need to look things up in other sources (which I hope the reader will be inspired to do, this is not social media, you can go wherever you want), one is out of the narrative of my writing anyway. So endnotes are exits, and escape hatches. This means that in the main text sometimes several lines of thought (or garden paths) occur simultaneously (using too many brackets and other punctuation marks) – this is on purpose, I am not writing a best-seller. This may seem lazy of me, but it actually takes more time and effort to write like this instead of a single-thread narrative. It is my expectation that the reader can speed up and skim if things seem less relevant, and focus with more rigour on the important bits.

I wrote this book from the stance of being in *awe* of the vast complexity of nature, from ecosystems to individual organisms, particularly the human mind and social behaviours. To do justice to this complexity (and awe), the structure of the book with all its many interrelationships might come across like the entangled structures found in nature. In the 1980s in the book *Mille Plateaux*, Gilles Deleuze and Félix Guattari introduced the rhizome as a more appropriate metaphor for the philosophy of complex ideas, rather than trees or roots. "We're tired of trees" they wrote, as reflected in taxonomy structures that "proceed by dichotomy", while the rhizome interconnectedness inspires different descriptions.[10] Mycelial networks, the underground existence of fungal hyphae, go even further. Recent research, as presented by Merlin Sheldrake in *Entangled Life* [2020], has shown that these intertwined communication networks stretch in some cases over kilometres, and operate in a mostly symbiotic relationship with the roots and trees of plants. These mycorrhizal networks, consisting of fungal hyphae and roots of trees and plants, not only exchange nutrients between each other, but also act as a messaging system – primarily to regulate the distribution of nutrients but of such a high sophistication that it is described as the 'wood wide web'. Indeed it resembles the decentralised structure of the World Wide Web and particularly the underlying Internet, but as Sheldrake points out, maybe the focus shouldn't be on the 'wood' but on the fungal underpinnings of the network – a 'myco-centric' perspective.[11]

OVERVIEW OF CHAPTERS

Dividing up the complex and deeply interrelated developments relevant for *Understanding Interaction* in chapters is inevitably a compromise. The historical development is not a linear story of technological progress and 'revolutions', but many slow (and eventually accelerating) processes, simultaneously developing technology, language, and culture. While there seems to be a chronological structure on the surface, in order to make the necessary connections and creating relevance, there are often links to the present and future (and what may seem like side-tracks) and even whiplash flash-forwards.

Likewise, the insights and knowledge drawn from the human sciences and communication theory are often interrelated in a way that there is no optimal order of covering the topics to get to the understanding of interaction. Throughout the book attempts are being made to link everything together, to come to a holistic understanding. This would also support reading the chapters or sections in a different order, and I have tried to include many cross-references throughout the text, to link different related sections and support independent reading (inevitably with some repeats).

Understanding Interaction Volume 1 – Evolution, Technology, Language, and Culture: Where Do We Come From and How Did We Get Here? is about evolution, technology, language, and culture. It has a **Preface** which introduces the book, including the structure and this overview of chapters. The preface is followed by chapters that lay out the field, and chapters that are engaged with the historical overview and reflection on the evolving and developing relationships between people, technology, culture, and language.

In **Chapter 1 Interacting**, the background of *Understanding Interaction* is presented, including definitions of interaction, the various terms used, and fields in this area. This is done through a historical overview of the different trends and concepts related to the study and design of tools and interfaces to interact with (particularly digital) technology since the mid-20th century. I sincerely hope that my decision to weave my personal trajectory of development into this overview hasn't hindered the message. My aim with this strategy was to ensure that most of it (but not all, of course) was experienced from the inside, personally felt as a close witness, participant, and perhaps even contributor. And while attempting to avoid over-subjectivity, I think this is preferred over detached objectivity in this instance. In many other instances throughout this book, where it is appropriate and even essential, of course objectivity has been supported.

The chapter also looks at education in this field, and reflects on experiences and approaches to teach understanding interaction. The aim of this book of course is that it can be useful in that context.

Drawing knowledge and insights from a large range of disciplines is essential for understanding interaction. Various strategies are discussed for combining disciplines, and in some ways rising above discipline boundaries.

The next part of the book is a historical reflection on evolution, technology, language, and culture (Chapters 2–6), looking at the historical developments from the Stone Age onwards, millions of years ago. This is when our ancestors started to make tools, and simultaneously developing our culture and language, a complex process of co-evolution still continuing. This relationship between technology and culture is a reciprocal connection, which can't be separated; some sections in this part of the book focus more on technology and others on culture. To link different related developments, a strictly chronological coverage of events is often not appropriate, so I often digress from the historical path through 'flash-forwards', particularly when it concerns developments and inventions that can be seen as the start of a long track of influence, lasting until the current time.

Chapter 2 *Evolving* presents the historical background of the relevant developments. The chapter introduces the 'technocultural periods', and a classification of technologies relevant for interaction. In the next chapters each of the technocultural periods is covered, roughly following a common categorisation in historical overviews: the toolmaking period, the settlement period, industrialisation, and the information period. These names are a compromise, a shorthand for a whole conglomerate of changes and developments in technology and culture. And although I am using the term 'periods', they only have a vaguely specific starting point in time, and with no end point; all these developments are still carrying on today. I deliberately avoid using the term 'revolution' (such as 'cognitive revolution' and 'industrial revolution') as upon close reflection it is clear that many developments are gradual and slowly progressing towards certain significant change – only retrospectively identified and therefore possibly somewhat arbitrarily chosen. Revolutions are often just oscillations towards a new equilibrium – and often not that much different from before. The focus is on the main trends, which often span large time periods – millions of years in the case of artefact making, hundreds of thousands of years in the case of language, and thousands of years in the case of mediated communication (from handwriting, to printing, to telephony, radio, TV, and the internet).

Chapter 3 *Creating*, looks at the 'tool period' starting from about three million years ago, drawing from insights in archaeology and palaeoanthropology. There has been a co-evolution of the physical and mental development of hominin species, not only through environmental factors but also through tool use, language, and culture. Four phases of tool making are identified: accidental or opportunistic appropriation of tools, dedicated tool making, tool designing, and compositing tool techniques. The chapter also covers the use of fire, and the evolutionary consequences of walking upright, such as the ability to use tools and develop more sophisticated vocal language including the use of symbols. Several ancestor species are briefly discussed, to illustrate the non-linear development leading to our species – not a hierarchical line but a meshed structure of interrelationships, of which many elements are still unknown.

In **Chapter 4** *Settling*, the 'settlement period' is presented, which started around 11,000 years ago, when people who until then had lived as mostly nomadic hunter-gatherers started to settle and practice agriculture and the domestication of animals. This led to the invention of various mechanical tools, new art forms, new cultural practices and societies, and writing systems using first mimetic, and later symbolic (abstract) signs. This deeply changed the way people related to their environments, which became a combination of natural and technological elements. The increasing size of settlements created pressure as well as opportunities, resulting in the development of hierarchies, specialisations, and other new social structures in societies.

At the end of this chapter, a critical reflection on this transition from nomadic to sedentary life is presented. This reflection compares traditional cultures with other contemporary civilisations, and what potentially can be learned from neglected insights and practices.

The next big change in human history, covered in **Chapter 5 *Industrialising*,** starts around 500 years ago with developments in scientific reasoning and discoveries, new technologies developed (mechanical, chemical, and later electrical), new attitudes towards religion and power structures emerging, the Renaissance development in the arts, and the Enlightenment. The most important development is that of industrialisation, as it profoundly changed the way people relate to technology. Previously, manufacturing was based on individual skilled handwork and craft, the 'cottage industry', but from the early 18th century onwards manufacturing became predominantly factory based. This had a profound impact on the structures of societies, including the military application of industrialised warfare. The increasing scale of this industrialised manufacture also has had an ongoing, accumulative negative effect on the natural environment.

The invention of the printing press with moveable type effectively facilitated the industrial manufacture of the printed book, a medium that had a big impact on society (including the Reformation, science, and art).

The technological categories are discussed, as well as a framework of technological functions. Just like the previous chapter, this chapter ends with a critical reflection on the notion of linear progress. As a complex conglomerate of changes and influences, industrialisation has not only brought prosperity to (parts of) humankind, it also deeply affected the natural environment, which in turn threatens our very existence.

Chapter 6 *Communicating* looks at communication, information, and representation. Just as the making of artefacts has a long history of development from the first stone tools to mechanical and electrical systems, the development of our languages and ways to communicate through media has a similarly long history.

This chapter discusses semiotics, the study of signs, how objects and ideas are represented. A sign 'stands for' something, it is a reference, a representation, not the thing itself. An object has a presence of itself (and can as such already have meaning and opportunities for application), and this physical presence can be *re*presented for instance in an image (mimicking the object, iconic representation), or in an abstract sign (in an arbitrary, learned relationship, symbolic representation). This is a continuum, a representation spectrum on which a sign can be defined in a purely abstract way or in a mimicking way with a strong likeliness, or anything in between. On this spectrum, various communication modes are defined (such as written language, drawing, speech). In semiotic terms, the object or concept, the *signified*, is represented in the *signifier* (the word or picture) or *representamen*, and how it is received, the *interpretant*. Meaning is not just something that is transmitted, meaning is usually established in the communication and interaction between participants, through physical objects and representation. The meaning established is further dependent on the knowledge and mind-set of the individuals involved, and on the context (*pragmatics*). There are also many forms of communication which are not about conveying meaning (such as phatic communication, small talk), including the non-verbal and contextual. All of this is often ill-supported in our current digital tools and social media.

Also covered in this chapter is information theory, including the notion of the *bit* as the unit of information. There is a hierarchy in meaning and usefulness of information, from raw data, organised data, information, knowledge, and insight, to wisdom. The chapter furthermore looks into the reciprocal relationship between language and thinking (linking to Chapter 10).

Connecting knowledge from semiotics, with its emphasis on visual and auditory communication between people, to theories about manipulation and interaction between people and physical objects and environments, this chapter briefly discusses the notion of affordances as an ecological concept. Affordances are properties of an object that can communicate the potential for use (intentionally or interpreted, appropriated), and therefore depend on the perceiver and context. This important topic is discussed in detail in Chapter 9 *Experiencing* which looks at human perception.

Volume 1 is rounded off with an **Interface**, an interlude consisting of a short recap, presenting preliminary frameworks, and creating a connection to Volume 2.

Volume 2 is about people, society, and design. This part of the book is about understanding people (Chapters 7–11), looking at how people's physical and mental abilities and activities are related to technological environments. It covers how we sense our environments through our senses, which we experience and become aware of through perceptual processes, we can respond, reason, and reflect on the world through our thinking, both cognitive rational processes as well as subliminal processes, and we can influence the world through our effectors such as speech and movement. This is where we see the communication modes presented in Chapter 6 in action. This order is a compromise of course, as there are many activities and behaviours that have a different order of processes.

Chapter 7 *Being*, introduces this part, and covers some key concepts and insights relevant for understanding interaction from psychology, physiology, anatomy, biology, and neurology.

We consist of a mechanical structure (our skeleton, tendons, cartilage), operated by muscles which are mechanical and chemical, controlled through electrical signals from the brain and further influenced by chemicals. There are many dynamic and communication systems in the body, electrical (neurons) as well as chemical (hormones). Further complexity comes from the interrelationships between these systems, which are much more elaborate than in human-made artefacts and technologies. Due to this, when technical systems are used as metaphors for living organisms it is often limiting understanding. This can be seen in the historical example of dualism, and later in behaviourism, and early cognitive science.

The chapter also covers the notion of neuroplasticity, the ability of the brain to change its neural connections to some extent, usually through training and conscious effort.

In **Chapter 8** *Feeling*, our senses are described, from receptors to neural connections with the brain. We are sensitive to many things in the environment, such as light (through the photoreceptors in the eyes), vibrating air (mechanoreceptors in the inner ear), physical objects, shapes, surface textures and forces (mechanoreceptors in the skin and inside the body), acceleration and orientation of the body (through mechanoreceptors in the balance organs in the head), smell and taste (chemoreceptors in the nose and mouth), temperature (thermoreceptors), and damage, which is sensed by free nerve endings and nociceptors.

There are many more sensory modalities than the proverbial 'five senses'. This chapter gives an overview of all the senses and presents a framework of human sensory modalities. There is a strong focus on the somatic (bodily) senses, as reflected in the title of this chapter, *Feeling*. The somatic senses are often ignored or underused in mainstream interactions with computers and other contemporary technology. Particularly, there is great potential for further addressing the human sense of touch (though the cutaneous sensitivity of the skin) and movement (through proprioceptors inside the body) in interaction, our tactual senses. These senses of touch and self-movement are called tactile and kinaesthesia respectively; the combination of the two is called haptic, all these have a passive (imposed) and active (self-obtained) mode. Collectively these sub-modalities are known as the tactual sense. Thorough knowledge on how these complex tactual senses work is essential as a basis for further applications in interaction. This knowledge is presented here, including a framework of all tactual sub-modalities based on the literature. Examples and applications of interface design and interaction research around these senses are covered, particularly of interfaces and devices with active haptic feedback.

Chapter 9 *Experiencing* looks at perception, with a phenomenological perspective (which has an emphasis on subjective experience), as well as an ecological approach (which emphasises perception as an activity rather than a passive process of receiving stimuli). This makes the ecological approach very relevant for understanding interaction, as it is about the notion that we perceive the world in a meaningful way because we interact with it. We often perceive objects and their properties and possibilities, including other entities in our environment that we interact with, immediately and directly without cognitive processing (direct perception), because of our active exploration and engagement. This direct perception is clearly shown in the notion of affordances (as mentioned in Chapter 6 on communication), through which we (like other animals) can directly perceive the potential for use of an object or other entity in our environment. The theory about affordances is a very useful and important concept for understanding interaction, though when approached in isolation not always fully understood. Furthermore, through its overuse and application in domains where other already suitable terms were available, it has lost a bit of its pertinence over the last decades. But we really need this term for understanding interaction, because of the emphasis on physically interacting with our environment. This *affordance of affordances* is argued by discussing the notion of affordances in the wider context in which it was conceived – that of direct perception and the ecological approach, as well as looking at the historical precedents (in Gestalt psychology and phenomenology), and in related contemporary extensions and insights such as enactivism and sensorimotor theory. These are all approaches that look at an organism in interaction with its environment, with all its richness and messiness, while in other approaches such as in early cognitive science and even stronger in behaviourism, the emphasis was on controlled studies in (by necessity) more static and artificial situations. In cognitive science, perception is often seen as a set of processes in the brain, less engaged with the environment, instead of perception as an experience of resonance with the world around us.

Instead of introducing new terms to cover the same concept (potentially including the same misconceptions) of affordances, such as 'signifiers' or 'feedforward', the *affordance of affordances* framework of terms is presented which extends the use of the term affordance to cover a range of variations and modulations which include further essential and more subtle elements in understanding interaction.

It is difficult, and impractical, to draw hard boundaries between sensation, as covered in Chapter 8, perception as covered in Chapter 9, and cognition which is the topic of **Chapter 10 *Thinking***. This chapter looks at the many mental faculties, including memory, processing of information, focus of attention, multitasking, and knowledge. Again, I have tried to capture and focus on the knowledge and insights in this discipline that is most relevant for understanding interaction, such as the recent emphasis on the importance of unconscious thinking. The term unconscious is not the most optimal, as it suggests a hierarchy (just like the terms subconscious and subliminal) which reflects the common way of thinking about thinking as a predominantly rational process. Unconsciousness is also associated with earlier mystifications of how our minds work, including ideas about the meaning of dreams. Research in the last decades, using various terms, has shown how much of our thinking actually happens, of course, not at the conscious, rational, explicit, serial, and relatively slow level, but at an unconscious, experiential, conceptual, implicit, massively parallel, and much faster level. This level was often dismissed or underrated in earlier cognitive models as largely emotional.

Indeed this makes the unconscious hard to research, the fact that emotions, feelings, biases, prejudice, and preconceptions also inhabit this space. However, in interactions with our environment it is this experiential part of us that does most of the work. All this is no surprise to professional designers; many good and appropriate designs successfully engage with the experiential mind, and can be interacted with without the need for much conscious effort. However, our digital tools seem predominantly designed for the rational mind, based on the profound misconception that we are rational beings all the time. We're only occasionally rational, when we really try. Moreover, often that which we think are our rational thought processes are actually strongly influenced by unconscious processes. This has to be taken into account in the design for interaction with technology.

Not only our thinking, but also a large part of our perception takes place unconsciously. This is by necessity, as the amount of information entering through the senses is vastly larger than the processing capacity of consciousness and cognition. Contemporary insights into the unconscious therefore also support the notion of direct perception (and affordances) as presented in Chapter 9.

This chapter also explores the notion of multiple intelligences or the modular mind, also in an evolutionary perspective, which is an attractive idea, but drawing these modules with too strict borders seems valid only in limited and clearly defined areas, and not particularly useful for understanding interaction and behaviours in complex situations, nor for guiding future developments.

After this exploration of recent literature and insights in the 'new unconscious' and the different ways of our thinking and being, the chapter presents a new framework that separates the rational, experiential/conceptual, emotional, as well as the automated, and autonomous parts of our being, in the context of understanding interaction.

The final part of the loops of interaction with the environment, after taking information in and processing it, is about our output modalities. **Chapter 11** *Expressing* covers how we engage with and influence the outside world. Again, there are no strict boundaries, and many internal processes take place before an expression of thinking, an activity, or an utterance, is presented. There are many ways we can express ourselves and influence the world around us, including modalities such as the voice, movement, manipulation, and in many modes – expressions can be abstract (in the symbolic mode, such as spoken language, writing, and sign language), mimetic (in the iconic mode, such as drawing and gesticulating), and manipulative (directly interacting with objects and the physical environment). Many gradations and combinations are possible; we generally express ourselves radically multimodally, in a range of simultaneous expressions of meaning and manipulation, from the explicit and conscious, to the implicit and unconscious. A distinction is made between voluntary and involuntary utterances, and while the former are most important for interaction, the latter can also play a role.

The chapter particularly explores expressive modalities not commonly seen as relevant for interacting with technology, by drawing inspiration from alternative and non-verbal modes of communication between people (and animals) such as smell, blushing, excretion, facial expression, posture, locomotion, proximity, and particularly in gestural communication including sign languages.

A further important category of output modalities is through technological means, mediated expression (either with technology, or with other people). These media and interfaces can either increase the reach and scope of an utterance (amplified voice, radio transmission, telephone, email), or even open up new expressive modalities by sensing changes in or on the human body that cannot be directly picked up by other humans, such as bio-electrical signals, brainwaves, muscle tension, or chemical changes. Examples are the typewriter or computer keyboard, touchscreen, lie-detector, iris or fingerprint scanners, musical instruments, mechanical tools, and utensils, etc.

The chapter also looks into the notion of the 'quantified self', the fitness and activity trackers that we use to create a dataset about our behaviours (this data is often monopolised by large tech companies, only superficially available for the individual). This 'datafication' of behaviour has a lot of further potential for extending interaction.

A framework of applicable output modalities for expression is presented, covering a range of modalities and modes, implicit and explicit, involuntary or under conscious control, and mediated. This can be used for expanding and exploring new ways of interacting with (and through) technology.

The final chapter of the book is **Chapter 12 *Understanding*,** which draws all of the topics explored together in a holistic vision of understanding and extending interaction. The chapter discusses ways to apply the knowledge in design and development of new technologies and interaction possibilities. The frameworks presented throughout the book are drawn together here, and as palettes of possibilities they can be connected to opportunities for design. A succinct overview is given of research methods in design for interaction, and approaches for developing new insights.

Contemporary industrial fabrication paradigms are discussed, involving instant manufacturing techniques such as computer-aided design, 3D printing, multi-axis milling machines, and laser cutters. These techniques open up opportunities for individual design and manufacturing, as opposed to the traditional and often unsatisfactory (inadequate and unsustainable) mass manufacturing. This is the *un*industrial revolution, with potential for democratisation of the manufacturing processes tailored to individual needs and desires, as well as creating a positive contribution to sustainable practices. This is illustrated with a range of prototypes, proof of concept demonstrators, and products-on-demand developed and tested in recent projects in interactivated stroke rehabilitation exercise tools.

Upon critical reflection, one would realise that currently many products and services of different brands and different times are often surprisingly similar; any changes and variations over time are mostly superficial or only aesthetic. This is the result of an incremental design approach in which innovation has only the slightest role. There is a need for revolutions in addition to evolutions. The approach of *radical redesign* is proposed, based on lateral thinking with a thorough knowledge of historical developments, as well as thoroughly understanding people's individual needs and abilities.

As many other researchers in the field of design for interaction have argued, there is an opportunity (and urgent need) to shift design and industrial practices from creating 'solutions' to creating tools, tools that are empowering people to create their own solutions and situations, reflecting their individual needs and wishes. With the most flexible and malleable set of technologies at humankind's disposal, this should be more possible than ever before.

Computer says '*yes*'.

NOTES

1. *Understanding Media – the extensions of man* [McLuhan 1964].
2. *Understanding Design – 175 reflections on being a designer* [Dorst 2003].
3. In: *Interaction with our Electronic Environment – an ecological approach to physical interface design* [Bongers 2004, p10].
4. The journal article 'Anthropomorphic Resonances – on the relationship between computer interfaces and the human form and motion' [Bongers 2013] in a special issue on Organic User Interfaces.
5. [Dix et al. 1998, pp56–60].
6. This development is described in the book *The Story of Writing – alphabets, hieroglyphs & pictograms* [Robinson 2007].
7. See the book by Frank Wilson, *The Hand – how it shapes the brain, language, and human culture* [1998].
8. Steven Mithen describes this development in his book *The Singing Neanderthals – the origins of music, language, mind and body* [2005].

9. A good overview is Yuval Noah Harari's book *Sapiens – a brief history of humankind* [2014], and by Vybarr Cregan-Reid in the book *Primate Change – how the world we made is remaking us* [2018] which reflects on how the four successive technocultural phases have influenced us physically and mentally.

10. Philosopher Gilles Deleuze and psychoanalyst Félix Guattari published *Mille Plateax* in 1980, and the translation by Brian Massumi was published in 1987 as *A Thousand Plateaus: capitalism and schizophrenia*. I used the Bloomsbury Revelations edition of 2013, this section is based on the introduction chapter on rhizomes and the full quote showing their post-structuralist tendencies is "We're tired of trees. We should stop believing in trees, roots, and radicles. They've made us suffer too much. All of arborescent culture is founded on them, from biology to linguistics." [1987, p15]. This is probably a bit unfair on trees, as they are actually very complex structures, and also as metaphors have a lot to offer as designer and data-visualisation expert Manual Lima shows in *The Book of Trees – visualising branches of knowledge* [2014] (thanks to Jos Mulder who gave this book to me).

11. Merlin Sheldrake *Entangled Life – how fungi make our worlds, change our minds, and shape our futures*. [2020, p177]. Sheldrake also points out that trees can sometimes fuse with one another, a process called 'inosculation' [2020, p275, n27].

Author

Bert Bongers is fascinated by the relationships between technology, people and nature, which he has been researching and exploring for over three decades. He has a mixed background in technology, human sciences and the arts, developed through education (PhD Human–Computer Interaction Design VU Amsterdam, MSc Ergonomics/HCI UCL London, BSc electrical and computer engineering) as well as practical experience in numerous R&D and design projects. In his work, he combines insights and experiences gained from musical instrument design, interactive architecture, video performances, and interface development for multimedia systems to establish frameworks and an ecological approach for interaction between people and technology.

He has worked at several prestigious art and music institutions in The Netherlands, at the Philips company, the University of Cambridge, Metrònom Gallery in Barcelona, and several universities in The Netherlands (TU Eindhoven, Vrije Universiteit Amsterdam, Hogeschool van Utrecht) in a range of disciplines such as design, architecture, media technology, communication, and computer science.

Since 2007, Bert Bongers is Associate Professor at the University of Technology Sydney (UTS) in the School of Design at the Faculty of Design, Architecture and Building, where he directed the Interactivation Studio from 2008-2017, a 100 sqm flexible and reconfigurable space and audiovisual infrastructure (multi-channel sounds system and video projections) for developing new paradigms for interacting with technology. At UTS, he has been teaching and developing subjects across a wide range of topics.

He has published widely in academic journals and conferences. His interactive audiovisual installations, projections, and performances have been presented at venues in Sydney (Vivid Festival, Powerhouse Museum, Cockatoo Island, Museum of Contemporary Art, Sydney Design Festival, NG Gallery, DAB Lab Gallery, and the Tin Sheds Gallery); Metrònom Gallery and the Mercat de les Flores in Barcelona; and STEIM and other venues in Amsterdam.

Chapter 1

Interacting

Overview of design and research for interaction

So, Understanding Interaction has many facets and draws on many disciplines.

To illustrate the multifaceted nature of *Understanding Interaction*, in this chapter I am describing my own path along a range of disciplines, fields, degrees, projects, jobs, explorations, accidents, experiences, and influences I have been lucky enough to have had in the last thirty-plus years. It is nothing more than that: an illustration. The core of this chapter is a reflective overview of the various fields and disciplines that have been explicitly concerned with design and research into the interaction between people and (particularly digital) technology in the last fifty years. The storyline of my own experiences is by no means a suggestion of the perfect path, as most of it happened without any particularly grand ambitions other than a drive to explore, learn, experience, and understand, in a number of situations that I have been lucky enough to be in. There never was a plan.

These personal experiences and other reflections may not be relevant for everyone, the reader is free to skip or skim – in fact, this is true for most of this book. This meandering is deliberate, to avoid a one-dimensional linear narrative. As mentioned in the *Preface*, it is for this reason the use of footnotes and endnotes was avoided, as they bifurcate the narrative resulting in multiple streams, instead I have tried to keep many facets close together. It is my expectation that in this structure, the reader can pick up multiple streams of narrative simultaneously. Reading is only on the surface a seemingly linear activity; just like everything else the human mind is involved in, it is in fact a parallel multifaceted game of establishing meaning, (re-)constructing knowledge and insight. This is backed up and discussed at several points in the book, taking inspiration and insights from a large range of disciplines. Examples are: traditional storytelling and memory spaces that support oral culture, as for instance Lynne Kelly describes her recent books *The Memory Code* [2016] and *Memory Craft* [2019]; the all-at-once representation which Marshall McLuhan aimed for when he studied the transition from oral to literal culture in *Understanding Media* and other publications in the 1960s illustrated with his own 'probes' or aphorisms;[1] the Cubist style in visual art such as the works of Pablo Picasso and

DOI: 10.1201/9781315373386-1

Juan Gris in the early 20th century, which combined multiple viewpoints in one image; the immediate communicative potential of memes;[2] and the recent insights from psychology and neurology in unconscious thinking as a massively parallel processing mind, as Daniel Goleman discusses in *Focus* [2013], Malcolm Gladwell presents in *Blink* [2005], and the work of Daniel Kahneman (with Amos Tversky) in *Thinking, Fast and Slow* [2011], to name a few (discussed in depth in Chapter 10, *Thinking*).

INTERACTION – A JOURNEY

Understanding Interaction is a dynamic, evolving landscape which in this chapter I will describe through my own journey, from which I have seen the landscape starting to be formed. Everyone has their own path of mixed professional and personal developments, but because this is my book, I am taking the liberty of sharing my life journey (or garden path) of explorations and experiences in fast-forward in this chapter, including mentioning many of the people I worked with or for – it is important to acknowledge all the help and inspiration I received and all the opportunities I had to learn from experienced people, pioneers and leaders, and many participants from new disciplines I was lucky enough to collaborate with. And while there are personal elements throughout this book, I have put a lot of effort in being objective in all areas that need a neutral, unbiased, and scientific approach. I have attempted to back all this up as much as possible with precise referencing, while avoiding 'reference dropping' or showing off (pretending to be erudite) – the aim was to support all that is presented with the proper references, acknowledging sources from a range of disciplines, and to offer the reader the opportunity to use these as starting points for further explorations. That's why they are in endnotes, as they facilitate a way out. Due to the breadth of the scope of this book, it is inevitable that some sources are still missed, either minor or crucial. I will keep (re-)searching!

Interaction craft and science

My personal path started on the technology side, already in a mix: my first degree was a combined engineering course (BSc in 'Technical Computer Science', which brought together software and hardware, and digital and analogue electronics), supplemented with a fair bit of hands-on mechanical tinkering through working in bicycles and motorcycle maintenance. During the BSc degree from late 1986 I worked for six months as a student intern on the design and development of electronic musical instruments at the Studio for Electro-Instrumental Music (STEIM) in Amsterdam, and later my final graduation project in 1989 at the Sonology Department of the Royal Conservatory of Music in The Hague. This led to a permanent position as 'musical engineer' based at Sonology and collaborating with

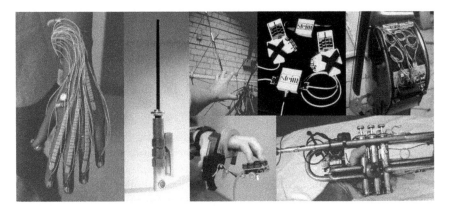

Figure 1.1 Electronic musical instruments (from left to right): The Lady's Glove (Laetitia Sonami, 1994), The Conductor (Michel Waisvisz, 1989), The Hands (Waisvisz 1988), The Web (Waisvisz 1992), The Meta-Trumpet (Jonathan Impett, 1993), Cello++ (Frances-Marie Uitti, 1999)

STEIM: a highly specialised (and rare) profession, working with top musicians, composers, and artists on very individual interfaces as shown in the images in Figure 1.1 and presented in several publications.[3]

Being immersed in musical culture was essential for understanding electronic musical instrument design, experiencing musical performances and rehearsals of such a high level of musicianship and instrument techniques, from early music and baroque, to classical, jazz, improvised, and contemporary composed music.

In this role I worked primarily as a craftsperson, an artisan, which is rare in digital technologies, certainly at that time.[4] In this position as a "luthier" of electronic instruments (as, among others, Nic Collins called my work[5]), I was very lucky to work with a range of extremely talented musicians, artists, and composers. In this decade of developing musical interfaces, I became particularly aware of the physical aspects of these electronic instruments, including haptic feedback, or rather the lack thereof – this formed the basis of further explorations.[6]

I became increasingly interested in the theoretical aspects of the interaction with technology and computer systems particularly, in addition to the craft and design aspects, an interest which I was able to develop in the mid-1990s through a position as research scientist at the Institute for Perception Research (known by its initialism in the Dutch Language IPO – Instituut voor Perceptie Onderzoek) in Eindhoven (NL), working for the consumer electronics company Philips on interfaces and multimodal interaction styles for networked home entertainment systems.[7]

All this work was mostly within the field of Human–Computer Interaction (HCI), as the entertainment systems but also the musical instruments I worked on earlier were effectively computer based. I read the first text book on HCI from cover to cover, which was written (as is often the case) by a multidisciplinary team with backgrounds in computer science and psychology [Dix et al. 1993]. But I felt the need for more knowledge about the physical aspects of the interaction, the importance of which I had learned through the work with electronic musical instruments, including explorations of active haptic feedback. In part this was what led me to Philips/IPO, as they at the time had a research track on tactual perception, and had developed a highly sophisticated force-feedback device[8] (a kind of motorised trackball, including software for designing the haptic feedback; this is discussed in detail in Chapter 8, in the sections on tactual perception). It was a great place to learn, and be part of teams that studied haptic feedback, and developed and tested new multimodal interaction styles.[9]

Around the same time (mid-1990s) I started working on spatial and physical interfaces for interactive and dynamic buildings with the architects Lars Spuybroek and Kas Oosterhuis.[10] This was for their two connected buildings of Water Pavilion (H2O Expo or Fresh Water Pavilion and Salt Water Pavilion respectively), and particularly with Kas Oosterhuis on other buildings such as the *TransPorts* interactive environment at the Architecture Biennale in Venice in 2000,[11] and *Muscle NSA* which was an interactive, real-time parametric physical structure using pneumatic actuators, for the Non-Standard Architectures exhibition in the Centre Pompidou in Paris in 2003/2004. Two of these projects were in 2014 partially recreated for the *Archaeology of the Digital – media and machines* exhibition at the Canadian Centre for Architecture in Montréal, curated by Greg Lynn.[12] These are shown in Figure 1.2.

Later I frequently collaborated with Oosterhuis at his Hyperbody research group at the TU Delft in Protospace, an interactive multi-user environment for real-time 3D CAD modelling for architectural design, developing a range of new physical interaction modules to control the parameters.[13] All this was a really interesting development from the intimate scale of interaction as explored as a luthier of electronic musical instruments, and

Figure 1.2 Sensor systems for audience interaction with buildings in architectural projects: Muscle NSA (Kas Oosterhuis/ONL) (left, inset sensor) and H20 Expo (Lars Spuybroek/NOX) (right)

already expanded in larger scale instruments such as The Soundnet and Global String,[14] to the monumental spatial scale of architecture. Many architectural practices already worked heavily with computer technology in their design processes (creating CAD models of future buildings), the work with these architects was about extending these capabilities with real-time manipulation of the parametric models and eventually spatial interaction in the actual buildings through large-scale pneumatic actuators.

In the late 1990s I worked as technical director at a postgraduate residential visual art academy, the Rijksakademie in Amsterdam. This allowed me to explore the role of interactive technologies in the visual arts, together with colleagues, artists, and participants. It was interesting to note that while many visual artists and sculptors had eagerly embraced video art as a medium, interactive media were far less accessible at the time. Their relationship with the technology was very different from musical practices. To create better support for this type of exploration at the Rijksakademie, together with the director Janwillem Schrofer I reorganised the workshop structure from artistic disciplines to broader applicable technology fields, and developed a large new digital and interactive media space, the Mediaship.

Interacting with mechanical technology – Ergonomics

In the late 1990s I pursued my focus on physical interaction and a more theoretical approach further through a Master of Science degree in Ergonomics at UCL in London, as there was no such degree in the Netherlands at the time. The UCL MSc degree combined HCI and the older discipline of ergonomics, and was led by Professor John Long at the time (it is now the UCL Interaction Centre, led by Professor Yvonne Rogers). In the MSc degree I learned that the discipline of ergonomics, or Human Factors as it is also known, particularly in the US, was founded in the 1940s. This actually struck me as odd, and somewhat limiting in scope – surely people were concerned about interaction before the 1940s? I was aware that people have been shaping technology to suit their purposes in previous times, such as the handles and levers to control steam streams, looms, printing presses, hand pumps, etc. This is why I extend the scope of ergonomics to all technologies developed by humans. This goes all the way back to the Stone Age, as I realised that we have been shaping materials and objects to suit our needs for millions of years. The first human-made artefacts, and for that matter the animal-made artefacts, show how technology is shaped to support its function and suit the needs and desires of the animal using it. This is indeed explored in the first part of this book, studying the co-evolution of physical tools, language, culture, and human capabilities.

Ergonomics is generally defined as the study, practice, and design of artefacts and environments for human work (from the ancient Greek words for work, *ergon*, and *nomos*, law, or measuring).[15] The term was first used in the Polish language in 1857 and introduced in English by Hywell Murell in 1949 with the definition "ergonomics is the scientific study of the relationship between man and his working environment",[16] or as Stephen Pheasant puts it "ergonomics is the science of fitting the job to the worker and the product to the user",[17] which reflects the way Étienne Grandjean defined it.[18] It includes anthropometrics, the science of body measurements. This emphasis on 'work' also struck me as rather limiting, particularly after a decade of my involvement in musical instrument design and making; it seemed appropriate to also include for instance tools for creative expression, play, and communication in the definition. Indeed Stephen Pheasant writes "the term 'work' may be applied to almost any planned or purposeful human activity, particularly if it involves a degree of effort of some sort". In these later developments the field was defined much wider than initially described by Murell and others, also including cognitive ergonomics.

In the 1970s the term Man–Machine Interaction (MMI) was used to take into account the interactive capabilities of the technologies developed (continuing the somewhat chauvinistic habit of the use of the term 'man',* but later the term was corrected to Human–Machine Interaction, HMI).[19] This eventually led to the field of study of Human–Computer Interaction in the 1970s and 1980s.

In the 1980s and 1990s other terms were used too, such as Cognitive Engineering (used for instance by cognitive psychologists such as Jens Rasmussen [1986] and Donald Norman [1986], Christopher Wickens's Engineering Psychology [1985, 1992], and Jakob Nielsen's Usability Engineering [1993], a term also used by John Long [Lim and Long 1994]). These terms show the strong links between cognitive science and engineering.

My final project for the MSc degree was an investigation into multitasking by including the tactual modality in the interaction, using techniques from electronic musical practice to investigate a low-level cognitive bottleneck (traditionally only studied between the modalities of visual and auditory, discussed in detail in Chapter 10 *Thinking*). As mentioned, the interest in tactual feedback started with musical instruments, because I had noticed that in electronic instruments any *active* tactual display of sound parameters was entirely absent, which can be seen as an omission when compared to acoustic instruments where the player can often not only hear but also feel the vibration (this is further discussed in Chapter 8 *Feeling*, in the sections on tactual perception).

Pursuing this interest further, after finishing my MSc degree I spent one year as a researcher at Cambridge University (in the Computer Lab and the Engineering Design Centre). Here I applied and extended my knowledge of

---

* In the vein of James Brown's famous song from 1966 *It's a Man's Man's Man's World* (adding "but it would mean nothing without a woman or a girl" as a kind of barely convincing disclaimer – great song though!).

tactual feedback and perception in interaction, developing haptic feedback interface paradigms for motion-impaired people.[20] In this project I first needed to demonstrate that our participants, all with Cerebral Palsy, were able to sense vibrotactile cues in spite of their impaired motor capabilities. After overcoming the initial resistance to the feasibility of applying haptic feedback, it later led to a further range of developments and experiments with haptic feedback systems by the group.[21]

Interactive art and architecture

From around 2000 I started to further explore a range of artistic expressions in interactivity, music, live video, and architecture in collaborative projects, in a new media laboratory I set up at the Metrònom Gallery in Barcelona (where I also curated and organised their yearly electronic music festival in 2001).

Working with the architecture collective Metapolis, I was teaching in their new MSc degree in Advanced Architecture at the Universitat Politècnica de Catalunya and participated in the development of the *Media House*.[22] This was an interactive architectural structure, with networked interactive technology developed by a team of MIT Media Lab led by Neil Gershenfeld. This was one of the first demonstrations of Gershenfeld's idea of 'Internet-0', a finely distributed decentralised network of sensors and actuators, based on his book *When Things Start to Think* [1999] and later leading to the notion of the Internet of Things [2004] (discussed in more detail below, and in Chapter 12).

I also worked on virtual textures or 'palpable pixels', still applying electronic music technologies in low-cost highly responsive experimental systems.[23]

The Video-Organ modular instrument for live audiovisual performances was developed to investigate how electronic musical instruments could be extended to include real-time video.[24] This work included the development of a new design space framework for physical interaction extending an earlier taxonomy I developed at Philips/IPO inspired by electronic musical instruments.[25] This Physical Interface Design Space is further discussed in Chapter 12.

This interest in combining and crossing disciplines, as an essential approach to explore new grounds, had by then led to the inception of a multidisciplinary performance group, the *Meta-Orchestra*, initially as a European project in 1999, and later developed in a range of projects directed by electronic and baroque trumpeter and composer Jonathan Impett, musician, composer, and video-artist Yolande Harris, and myself.[26] This ensemble, in varying permutations, brought together performers and practitioners from a range of backgrounds (including music, dance, design, architecture, video, sound art, and theatre) in a live performance setting. The other reason for using the term 'meta' was that all performers were connected in a real-time A/V and data network (Ethernet based) which meant that the participants could influence each other's systems in predefined ways – for instance, one of the sensors on an extended trombone instrument could affect a parameter of Harris's video instrument, or a tracked gesture of a dancer could influence a musical parameter in Impett's *Meta-Trumpet* AI-based algorithmic composition system [1994, 1996]. In Meta-Orchestra performances in various locations with different combinations of participants and conceptual approaches, large spaces and new presentation formats were explored not only in concert venues but also in disused industrial spaces.

A number of lessons were learned in this project, for instance that bringing multiple disciplines together didn't necessarily lead to a synthesis – a unified approach had to be developed, with a common language, which takes time. Mere juxtaposing people's individual disciplinary approaches and practices would generally not lead to a whole. First of all, we needed people who were already able or at least willing to connect with other disciplines and practices. Furthermore, the networking of the instruments turned out to have the potential we envisioned, to not just connect people's data but actually foster and support crossdisciplinary communications. In other projects where we combined live video with music without the networked connection between performers, there seemed to be a general tendency for musicians to ignore the video images (in some instances, they would even explicitly state this habit). In the Meta-Orchestra when the performers were able to actively control and connect to each other's set-up they were drawn into engaging with other modes of expression. A rewarding interaction can lead to increased engagement.

Another means of fostering a connection between image and sound was the use of Harris's graphic scores, which we extended through interactive video projections.[27] We also developed a portable videoprojection instrument, powered by a large battery, in order to 'liberate' the projector and be able to take it for videowalks.

INTERACTING WITH COMPUTERS

The computer – in all its forms: personal computer, supercomputer, digital sound processor, networked, embedded, smartphone, tablet – is the most powerful technology humankind has ever developed, and offers the greatest opportunities for rich interactions. As mentioned above, previous technologies from the first stone tools to mechanical and electronic systems have increasingly offered interaction potential, including manipulation and operation. This development of interaction starts with the earliest physical tools, then the mechanical systems (first passive, later active machines), later electric, analogue electronic systems, and finally digital systems. The approach of *Understanding Interaction* is to look at all interactions, extending the scope of the fields of study of the interaction between people and computers to all technology.

The key aspect that makes the computer different from all other technologies is not only that it can store and process information, but that it can be programmed (through software) to perform different functions and carry out different processes and algorithms. This was the case with the earliest active mechanical systems during the industrialisations (such as the loom, and Charles Babbage's Difference Engine from around 1830), later with the analogue electronic computers in the first half of the 20th century, and particularly the computers based on digital electronic technology from the mid-20th century. The digital computer has become increasingly intertwined with our everyday lives, becoming omnipresent, interconnected, and miniaturised, embedded into other technologies, everywhere in the environment. This omnipresence has increased the need for appropriate interfaces. Where earlier technologies due to their physical presence expose some of their workings and operations, and in some instances can be controlled by directly manipulating the technology (for instance mechanical systems, and even to an extent analogue electronic circuits can be influenced directly by touching them), such direct interaction is entirely absent with digital systems, unless an interface is present. This interface needs to be designed and implemented – only then are we able to interact with the digital system. And while the computer technology has become increasingly powerful, the practice of miniaturisation makes it even less visible, audible, tangible. This has been a trend from the first computers in the 1940s that filled a whole room, to the computers in the 1970s when a mainframe computer was the size of a large cabinet, to the desktop and laptop computers in the 1990s. Following Moore's Law, first formulated in the 1960s by Intel founder Gordon Moore,[28] about the exponential increasing processing power and memory storage due to the number of transistors that fit in a chip that was doubling every 18 months or so, computer technology has been shrinking to the point that there was less room for the interface. This can be seen with the mobile phones in the 1990s, from a brick size to the razor thin handheld devices.* At the same time, more functionality was afforded by this miniaturisation, making the first 'smartphones' hard to use, until Apple introduced the multi-touch screen in the first iPhone in 2007 which set the trend for all smartphones since. Perhaps it relied a bit too much on the

---

* The famous *Zoolander* mini-phone is a little flip top device that Derek Zoolander (Ben Stiller) could barely hold in his hand, envisioned this scenario. It was so hilarious in the 2001 movie that it made its reprise in the 2015 sequel (outperforming the iPhone of Valentina Valencia (Penélope Cruz)). Examples of this type of exaggerated miniaturisation are not restricted to movies; when visiting the SEG electronics market in Shenzhen in early 2019 I bought a tiny mobile 'feature phone', made by the company L8star. This BM10 is modelled on the iconic Nokia 3310 (itself recently reissued, originally from the year 2000, also immortalised in concrete in one of Will Coles' physical graffiti objects from around 2010 – see the section on 'twisted affordances' in Chapter 9), but shrunk to almost a quarter of the size and it actually works! It has a very effective OLED display, and many other technical features not included in the (bigger) original. Mine is orange, but there was a blue (steel?) model too, and a later model which is a flip top. I also bought a mini-'iPhone' (Android) by Uniscope, about half the size of an iPhone plus – this one is a bit hard to use too but very funny, and actually an important statement about the availability of devices of different sizes for different purposes.

touch screen, some functions could be much more appropriately supported with other technologies (the LED indicator for message status, an analogue volume dial, buttons for media playback). But the iPhone development did halt the miniaturisation – the devices only became thinner, and eventually these phones actually started to become bigger again!

Due to the miniaturisation of digital technology, and because of the increased networking and interconnectedness, it is omnipresent without necessarily being physically present, which is discussed in a section below, on the 'electronic ecology'. But these two opposing tendencies of the technology becoming increasingly powerful, and at the same time disappearing, means that the need for appropriate and well-designed interfaces is stronger than ever.

Early visions of computer interaction

It is interesting to note that many of the often quoted early and indeed seminal papers and research projects in computer technology focussed on the interfaces and interaction, rather than on the underlying technology. The research explored the potential for interaction, often explicitly as a means of extending human mental capabilities. For instance the 1945 paper *As We May Think* by Vannevar Bush, who worked on the development of analogue computers (and also worked on the Manhattan Project that developed the first atomic bombs) imagined his 'memex' system for 'memory augmentation'. The memex was supposed to use information stored in microfilm (based on chemical and optical technology of photography at the time) available through mechanical contraptions and controlled through buttons and levers, and also envisioned speech recognition.[29]

Another positive side effect of the WWII efforts was the support of the research of Norbert Wiener, who developed the notion of cybernetics initially through his research on missile guidance systems (and, like Vannevar Bush, in the 1950s became more critical of military applications of technology). Wiener combined knowledge from biology with engineering and mathematics in his first book *Cybernetics – control and communication in the animal and the machine* published in 1948.[30] Cybernetics, a term Wiener derived from the Greek word for steering or governing, was about control theory and established a basis for the study of the interaction between people and computers by comparing their inner control systems. People and other animals guide and articulate their movements and physical actions through feedback loops, and so did the improved version of the steam engine that used a 'governor' (a spinning mechanism with moving weights) to keep the speed constant through a negative feedback loop.[31]

Going beyond mere interaction and advocating a symbiotic relationship was an inspiring publication by J. C. R. Licklider, *Man-Computer Symbiosis* [1960].[32] Licklider had a mixed background in engineering and behavioural sciences and specialised in speech and sound technologies.[33] Although there is that gender bias again (here is a quote from the summary at the beginning of the article: "to enable men [sic] and computers to cooperate in making decisions and controlling complex situations"), it was a visionary publication full of important ideas around AI, networks (Licklider later led the research at DARPA, the Defense Advanced Research Projects Agency, which led to the introduction of the Internet), drawing interfaces and handwriting recognition, wall displays, and speech input and output. A symbiotic relationship aims for mutual benefit in the interaction between people and technology, and is still a long way off in many contemporary systems, even though computer technology plays such a large role in our intellectual development.

One of the people who was inspired by these ideas, and further developed and actually implemented them, was Douglas Engelbart later in the 1960s.[34] In Engelbart's work too there is a strong focus on supporting human intellect, facilitating the emergence of collective intelligence, and extending our abilities through technology. This is reflected in the term Engelbart uses, "augmenting human intellect", and with his team he developed a really broad vision of the future of computer interaction. This included inventing and applying new input devices such as the mouse (with electrical engineer Bill English), initially a wooden box with two rotating wheels connected to potentiometers.[35] While the mouse eventually became a mainstream input device, they also report a now largely forgotten input technique of a knee-control device mounted under the table with left-right and up-down degree-of-freedom mapped to the movement on the screen of a cursor (or pointer, which they called 'bug' at the time). In a later demo by Engelbart and his team of the Augmentation Research Center at the Stanford Research Institute, the whole oN-Line System (NLS) was presented live for an audience at a computer conference in 1968 in San Francisco and linking to their lab at Stanford University in Menlo Park. They basically showed the whole Human–Computer Interaction paradigm as it would become mainstream decades later.[36] The video of the demo shows the participants collaboratively working on text documents and computer code, graphic displays, hypertext, windows, the mouse, and pointer (in the demo called 'tracking spot'). The demo was also showing elements that seem to me really effective and appropriate but unfortunately never became mainstream, such as a chording keyboard for faster text input with one hand. Just like the knee-control device, there are certainly advantages in these techniques – though obviously also some issues, for instance the knee-device needs to be firmly attached to the bottom of the working surface which is not always possible, and chording keyboards have a learning curve.

There were other pioneers in the 1960s. For instance Ivan Sutherland, who invented the 'sketchpad', the first computer graphic drawing system which allowed people to communicate with the computer through drawings.[37] Another important pioneer was Ted Nelson, who worked on complex and interrelated information networks and coined the term 'hypertext'.[38] Nelson, with a background in philosophy and sociology, envisioned something much more sophisticated than even current interactive media, much more approached from a literature perspective. For instance, the hyperlinks as proposed by Nelson were meant to be bidirectional, and he made other proposals which allowed for much more elaborate and complex networks of connected content, potentially supporting deeper understanding, and creation of content. He remained a strong voice in challenging the computer interaction paradigms.[39]

In the early 1970s many of Engelbart's ideas were picked up by Xerox PARC (Palo Alto Research Centre), including many members of his Augmentation Research Center at SRI such as Bill English, where they created the first computer system with a graphical user interface (GUI).[40] Their mouse-pointing device was an improved version of Engelbart and English's design, now using a roller ball instead of the wheels, which improved diagonal movements. The Xerox team involved Alan Kay, who implemented knowledge from developmental psychology, particularly the ideas of Jerome Bruner.[41] Like Jean Piaget earlier, Bruner realised people (and children) learn in stages, which he called **enactive** (manipulating objects, in Piaget's terminology this was the kinaesthetic phase), **iconic** (using pictorial, mimetic representations of objects and concepts, using metaphors such as the desktop and trash can), and eventually **symbolic** (purely abstract languages, using symbols which have no resemblance to the objects or concepts they refer to, based on arbitrary and learned conventions). (These stages or categories are discussed in further detail in Chapter 6, on information, communication, and semiotics.). In the GUI paradigm all these stages are applied and presented simultaneously: with the mouse and the pointer graphical representations can be manipulated (by pointing at them), there are icons (which pictorially represent files and other entities in the system, through metaphors), and symbolic (textual) representation. It is a good (and unfortunately, rare) example of how design of new products, systems, and paradigms is best approached through a firm basis in scientific research and knowledge, and a thorough understanding of how people do things through studying their activities, gathering insight in people's wishes, needs, and desires, at any stage of the design process.

Although impressive, the Xerox Alto system that was brought out in 1973, and the later Xerox Star as a commercial product in 1981,[42] were expensive and had a few other drawbacks that made them hard to market. The vision of the Xerox company as a developer and manufacturer of office equipment such as photocopiers was that of the 'office of the future'. The laser printer Xerox invented in the process became a commercial success, as we are all aware of,[43] and the company invented Ethernet as part of this office of future vision. The well-known story is that after seeing a demo in 1979, Steve Jobs decided that Apple should develop this concept further and make it popular for a wider audience. One of Apple's first challenges was to make the system more streamlined and cheaper, which included developing a mouse-pointing device that was affordable, and many other essential developments.[44] The first attempt of Apple was the Lisa, a computer with a graphical user interface for the professional market, which was not a success because of its high price (still) in comparison to the performance. Apple's other project developed simultaneously; the Macintosh, which was developed by a team started by Jef Raskin and later involving Steve Jobs, became a commercial as well as conceptual success after its introduction in 1984. By now the GUI had a fully-fledged implementation of the windows, icons, menus, and pointers (WIMP) paradigm, particularly through the contributions of Bill Atkinson who developed the pull-down menus, overlapping windows, 3D graphics, and halftone images.[45]

Microsoft had been working on graphical interfaces as well, with their Windows system which only became more popular in the early 1990s in version 3.1, and eventually particularly with Windows 95 (indeed, in 1995 it would have been version 4 but that didn't look good next to Apple's operating system which was already in version 7 by then, since 1991, and would introduce version 8 in 1997).

Interface design

There have been many terms and many subfields introduced over time within the field of Human–Computer Interaction, which cover the study and design interfaces for interactive computer systems. These terms and their meanings shift and change over time.

In the late 1980s and the early 1990s the term 'interface design' was in use, which seemed appropriate to me as that was what I felt I was doing at the time (after all, the interface is what facilitates the interaction, such as the musical instruments developed); it has a focus on the physical entity as well as the interactions that emerge through it. In that time the term had this broader view, for instance one of the earliest books on HCI, assembling a number of writings of key researchers in the late 1980s, was called *The Art of Human–Computer Interface Design*, edited by Brenda Laurel [1990], Ben Schneiderman's book *Designing the User Interface* – the first edition came out in the late 1980s [1986], now in its sixth edition in 2017, and other books in the early 1990s such as Lon Barfield's *The User Interface*,[46] Deborah Hix and Rex Hartson's *Developing User Interfaces*,[47] the first edition of Alan Cooper's *About Face: the essentials of user interface design*,[48] and another book to emphasise the importance of the people who actually use the technology, the very inspirational *The Humane Interface* by Jef Raskin.[49]

But the term 'interface' is limited nowadays by the view of some that the interface is a web site, or a set of buttons, a focus on the actual interface and not the dynamic and interactive aspects of it. Furthermore it is also limiting the scope to the design of the interface and not the underlying system, the interface as a tagged-on element which as we know is not the right approach. As mentioned above, as I wrote in an earlier book, comparing it with the role of the human face (skin, eyes, nose, ears), "the (beauty of the) interface is more than skin deep."[50]

User-centred design

To emphasise the importance of the people using the technology, rather than designing the technology itself, the term User-Centred Design was introduced in the mid-1980s by among others Donald Norman in *User Centered System Design – new perspectives on Human–Computer Interaction* [Norman and Draper (eds.) 1986], and in Chapter 7 of his seminal book *The Psychology of Everyday Things* written in the late 1980s.[51] The term User-Centred Design is still widely used, as well as the term Human-Centred Design.

Strictly speaking though, design is always about the user, or the human, or at least it should be – if isn't about the user, it isn't design! In that sense, the term 'user-centred design' is a pleonasm. But the term is clearly a response to many technology-driven developments particularly at the time, and it is unfortunately necessary to re-emphasise the importance of the design focus on the people who use the technology. This is reflected in the title of Alan Cooper's book *About Face*, stressing the importance for developers of technologies to turn around, away from the technology, do an 'about face' (turn around) and look at the people who use it. As stated above, an interface by definition has to face two ways, connecting the technology with people. Otherwise, it would be Inhuman-Computer Interaction! Or Humanless-Computer Interaction. There are indeed computer applications which have little to do with humans (automated control systems in factories for instance), but most computer systems are operated and manipulated by people, if not as the final purpose of the system then at least in co-existence.

The only time I use the term 'user' is when I explain why I don't like to use the term 'user'. I prefer terms such as 'audience' (because this has the connotation of a performance; it emphasises the responsibility to create worthwhile experiences) or just 'people' – nowadays we all use technology, in whatever form or complexity, for whatever purpose. The term 'user', as we know, was always an abstraction in HCI and particularly computer science and software engineering, and by many in these technology disciplines seen as something not necessarily crucial to the core of the work, the machine. The 'user' as a passive participant in a one-way activity, with the term 'end user' as the ultimate insult: the problematically irrational, unpredictable, and erratic actor in the whole technology show. And only in the English language is it possible that phonetically the word 'user' is very similar to 'nuisance' (perhaps not a coincidence, convenient even as it reflects the disdain that some technology developers have for other people). The ultimate responsibility of any technology is that it is satisfying, useful, and appropriate for people, and not to treat people as an abstraction, irrelevant for the technology itself, and therefore allowed to be conveniently ignored.

Human–computer interaction and interaction design

The term Human–Computer Interaction (HCI) has been in use since the 1980s or even the 1970s. Actually at first it was known as Computer-Human Interaction (CHI), a term which seems to place the technology first, not the human, and the acronym lives on in the biggest yearly HCI conference, alternating between the US and locations in other countries, 'the CHI' (pronounced as 'kai').

Looking for a term that appropriately described the design discipline related to computer interfaces, Bill Verplank and Bill Moggridge of the product design company Ideo came up with the term Interaction Design in the 1980s, resembling the name of the more established discipline of Industrial Design.[52] Though we don't really design the interactions themselves, but rather the ways that interactions can emerge, it is after all a dynamic process and depends on many factors when in use. The term Design for Interaction is sometimes used, which might be more accurate, Interaction Design works well too and is the most widely known. However, Moggridge's book with reflections and interviews with the key pioneers in the field of HCI since the 1970s is called *Designing Interactions* [2007].

One of the first Interaction Design degrees in the world was called 'Computer Related Design', an MA degree founded by Gillian Crampton Smith at the Royal College of Art (RCA) in London in the late 1980s, with Bill Gaver as one of the teachers who invited me for some visits and deliver a lecture in in 1997. It was very inspiring to meet the students in their studio, and some later came to the Rijksakademie (see above) as interns. Another early course was at the Utrecht School of the Arts (HKU) in the Netherlands, which started a bachelor degree in Interaction Design around 1990, led by Dick Rijken (who later became the director of STEIM).

Human–Computer Interaction can be seen as an umbrella term, the broadest term encompassing design, research, and study of interaction, and the later term Interaction Design as the applied field of study and development. This was reflected in the titles and scope of several main text books in various editions such as *Human–Computer Interaction* by Alan Dix et al., who in 1993 defined HCI as "the study of people, computer technology and the ways these influence each other",[53] and *Interaction Design – beyond human-computer interaction* by Helen Sharp, Yvonne Rogers, and Jenny Preece who define interaction design as "designing interactive products to support the way people communicate and interact in their everyday and working lives" (since the second edition in 2007).[54] In the more encyclopaedic *Human–Computer Interaction Handbook – fundamentals, evolving technologies, and emerging applications*, edited by Julie Jacko [2012],

Jonathan Grudin in the first chapter *A Moving Target: the evolution of HCI* gives an overview of the many different definitions (and other historical aspects) of HCI and related terms and how they shifted over time.[55] Ben Schneiderman in the foreword of the HCI Handbook proposes a distinction between micro-HCI (using established practices, heuristics, and approaches) and macro-HCI for "exploring new design territories".[56]

Software engineer and designer Alan Cooper and his co-authors in the 4th edition of *About Face* in 2014 define it as "the practice of designing interactive digital products, environments, systems and services", extending this with emphasising that it is about designing behaviour.[57] Cooper means the behaviour of the system, but of course this cannot be isolated from the behaviours of people. And, just like 'interaction' cannot be 'designed' but only 'designed for', behaviour cannot be designed, but only designed for, by creating opportunities for behaviours to emerge in appropriate ways (that is, preferably, not enforced). Cooper in his books gives many examples of how badly designed interfaces lead to inappropriate interactions and behaviours, as such reflected in the great title of his book intended for businesses, *The Inmates are Running the Asylum*.[58] Cooper makes the point that in the 1990s, software was often not developed with the 'user' in mind, that it was programmed for other programmers rather than 'designed' to be used by other people. He writes about how this can be improved by involving the people who actually need and use the products and systems in the design process.

Definitions of HCI and ID have been extended over time (the definition in the first edition of the textbook by Sharp, Rogers, and Preece was just "designing interactive products to support people in their everyday and working lives"),[59] often reflecting the backgrounds of those involved. For instance, influential designer Jon Kolko in *Thoughts on Interaction Design* in 2011 defines it as "the creation of a dialogue between a person and a product, system, or service. This dialogue is both physical and emotional in nature and is manifested in the interplay between form, function, and technology as experienced over time."[60]

In *The Encyclopedia of Human–Computer Interaction* of the Interaction Design Foundation published in 2014, Jonas Löwgren in the chapter *Interaction Design – brief intro* describes it as follows: "interaction design can be understood as shaping digital things for people's use".[61] John Carroll in the same e-book in the chapter *Human Computer Interaction – brief intro* reflects how the field has developed over time: "HCI now aggregates a collection of semi-autonomous fields of research and practice in human-centered informatics."[62] Initially HCI was mostly drawing expertise and approaches from cognitive psychology and computer science, but it has developed further and became a broader discipline by incorporating a wider range of disciplines, as mentioned in the *Preface* (shown in Figure 0.2).

Reflecting the complexity and multidisciplinary nature of the field of HCI, several of the important HCI publications and reference books were conceived by a team of authors, from different backgrounds (including the later versions of influential books by previously single authors, such as Ben Schneiderman, Alan Cooper, and Christopher Wickens [2013], all include a team of co-authors in their most recent editions).

Specialisations and subfields in interaction

There are many subfields of HCI and ID that have emerged over the years, focussing on particular target audiences or specific application domains, such as children, robots, animals, health, military, agriculture, music, art, games, entertainment, and automotive applications. This shows a much broader engagement than office productivity, striving for accessibility for all. There are also subfields (and associated journals and conferences) that explore particular interaction paradigms, such as multimodality, mobile, speech based, gestural interaction, wearables, and ones that focus on physical interaction. From the mid-1990s, examples are **Graspable Interfaces,** as Bill Buxton and George Fitzmaurice called it,[63] and **Tangible User Interfaces** (**TUI**) by Hiroshi Ishii and Brygg Ulmer.[64] This led to the **Tangible and Embedded Interaction** (**TEI**) conferences, held yearly since 2007. Since 2011, the word 'Embodied' has been added to the title of the TEI conferences, reflecting the notion as formulated by Paul Dourish in *Where the Action is – the foundations of embodied interaction*, which also gives a good overview of the many developments in HCI in the 1990s.[65] In design for interaction, 'embodiment' can have two meanings (both covered by TEI), one is the embodiment of meaning in an object (representing meaning through a physical object), the other is that the interaction involves the body of a person interacting with it, such as through dance or other movement.

The **New Interfaces for Musical Expression** (**NIME**) conferences are a platform for research and development of electronic musical instruments, founded in 2000 at a Workshop at the CHI conference. NIME is involving among others many old friends from my STEIM/Sonology years, and in 2010 we hosted the conference at UTS. I called this one NIME++, to accentuate a mission to extend the field beyond musical instruments, to more generally include interfaces for live performance and real-time generation of video material[66].

DEVELOPING DEFINITIONS

In summary, overall Human–Computer Interaction (HCI) can be seen as the more scientific study of interaction potential, while Interaction Design (ID or sometimes IxD), can be seen as the more applied field, but both are strongly research based.

However, Interaction Design nowadays means different things to different people, and has become increasingly narrow in focus for many, who see Interaction Design as mostly concerned with designing online or other screen-based experiences. This has led to the introduction of a range of new terms in the last few decades, each with their own emphasis, such as **Computer Supported Cooperative Work** (focussed on group cooperation and collaborative tools), **Participatory Design** and **Collaborative Design** (approaches which involve in the design process the actual people who will eventually use the systems), **Service Design** (particularly online systems that provide a service to people), and **Experience Design** (which has an emphasis on not just the functionality of a system). Experience Design is an approach that could draw inspiration from the disciplines that traditionally have been concerned with creating experiences, such as literature, cinema, music, or theatre (Brenda Laurel explored this in her book *Computers as Theatre* [1991], using stage performance and drama practices as a metaphor to stress the importance of audience experiences).

Interactivation

With so many different names for the same thing, or the same name for so many different things, it seemed necessary to introduce new terms to emphasise what I thought was most important. Since around 2000 I started using the term 'interactivation' as I realised the relevance of existing artefacts and architectures, and the related design disciplines, and the 'interactivating' of these. I first used the term in the architectural design context, working with Kas Oosterhuis and the Hyperbody research group at the Faculty of Architecture at the TU Delft, and the Metapolis architecture collective in Barcelona, in workshops and publications on "interactivating spaces".[67]

Of course 'interactivating', the process and practice of making things interactive, is a means, not a goal in itself. The goal is to open up technology, to give people access to its functions, and empower them to create, express their intentions, and carry out tasks. The term Interactivation is appropriate because it puts emphasis on *activity*, rather than on passive processes, and even the notion of *interactivism* – because I think it is really problematic if the interactivation potential of objects and environments is unfulfilled; there are huge opportunities here and we need to behave as interactivists to develop these.

The electronic ecology

Technological environments also have become increasingly interactive. With the earlier mentioned tendency of miniaturisation and networking of technologies, this leads to an increasing complexity and at the same time decreased visibility, or rather, a decreased interactibility. The need for a well-designed, appropriate, and insightful interface is stronger than ever.

Within HCI there are several terms that describe the research areas around interaction and distributed systems. Currently popular is the term **Internet of Things (IoT)**. Since the 1990s it was called **Ubiquitous Computing**, which has a bit of a sinister ring to it (which might be considered appropriate actually) or the shorter, more catchy (and less intimidating) **UbiComp**. Actually science fiction and technology writer Bruce Sterling in the early 2000s proposed the acronym as a more comprehensive version: "Ubicomp is what ubiquitous computing might look like after sinking deeply into the structure of future everyday life."[68]

Ubiquitous Computing is about the notion that computers have become miniaturised and networked, embedded in our environment. This completely changes how we approach design for interaction with these environments from the earlier situation of the computer as a single device. The notion of Ubiquitous Computing was first proposed in research projects at Xerox PARC by Mark Weiser and his colleagues and presented in an article in the Scientific American, *The Computer for the 21st Century* [1991]. Weiser would emphasise the importance of the computer technology blending in with the background, facilitating a peripheral interaction rather than always claiming the focus of attention, particularly when it doesn't need to be. Weiser's ideas, including his later emphasis on 'Calm Technology', have inspired many positive developments.

Since the 1990s, particularly in software engineering, the term **Pervasive Computing** was used, also potentially sounding a bit sinister.

The AT&T Labs in Cambridge, UK (previously a lab of Olivetti Research) since the late 1990s used the term **Sentient Computing**.[69] This sounds a lot friendlier, the word sentience emphasises the intention that the technology can be more aware of the people in the surroundings and enable access.*

Emphasising the fact that the computing technology is seemingly vanishing, are terms like the **Disappearing Computer** (a European project from the 1990s), or the title of one of Donald Norman's books *The Invisible Computer*,[70] and a book edited by Peter Denning, *The Invisible Future* with a collection of insightful essays from a wide range of researchers, scientists, writers, and thinkers.[71]

As the actual sensors and circuits have become in the micrometre scale since integrated circuits were invented, in the early 1990s the notion of **Smart Dust** explored how electronics could integrate in objects and environments, using MEMS (Microelectromechanical Systems such as accelerometers) and RFID (Radio Frequency Identity) chips, as a precursor of IoT. The tantalising idea was that the circuits would be so small that they could just be scattered around the environment and become a self-organising ubiquitous

---

* However, the word 'sentinel' means a watchful entity, like a guard, and also the name of the vicious robots that attack the *Nebuchadnezzar* in The Matrix movies. Entities such as Agent Smith are described as 'sentient' computer programmes'.

network. Although the chip-based components are on the micrometre scale (10^{-6} or a millionth of a metre), the actual circuits are at least on the millimetre (10^{-3} or a thousandth of a metre) or centimetre scale – still impressively small. (Nanotechnology is a hundred to a thousand times smaller, a nanometre is 10^{-9} or a billionth of a metre, the scale of individual atoms and molecules.*)

Meanwhile in the real world, the 'smart home' is becoming increasingly mainstream nowadays, the first work in this area was known as 'domotica' in the 1980s. Philips introduced the term **Ambient Intelligence**, which has an emphasis on the spatial and the peripheral, and particularly focussed on people's environments for living (the home) and work (the office).[72] Philips established their HomeLab in the late 1990s, extending the work I was involved in when I worked for Philips in the mid-1990s.[73] In the HomeLab, a house-sized experimental environment that Philips opened in 2002, researchers developed many scenarios and tests of how people could interact with ambient intelligence in the house.[74] It was later extended with more spaces for other scenarios (such as an office) in the Experience Lab in 2006.[75]

Things That Think is the term used by Neil Gershenfeld from the MIT's Center for Bits and Atoms (who is advocating 'personal fabrication', also discussed towards the end of Chapter 5 *Industrialising* and Chapter 12 *Understanding*). Gershenfeld wrote a book *When Things Start to Think* [1999], part of the vision of connecting things, the **Internet of Things (IoT)**, which claimed "The principles that gave rise to the Internet are now leading to a new kind of network of everyday devices, an 'Internet-0'."[76] As mentioned above, this idea was demonstrated by Gershenfeld in the Metapolis Media House project. This is the very inspiring notion of all devices and objects in our environment becoming networked, allowing for instance a dynamic configuration of lights and switches in our surroundings (now quite common with systems such as the Philips Hue, Apple HomeKit, Ikea's Trådfri range of smart lighting devices, etc.).

With these examples and short historical overview I am trying to show that the current trend of IoT, often called an 'emerging technology', has actually been around for three decades now. Similarly, other 'emerging technologies' such as VR and AR (Virtual Reality and Augmented Reality, also known as mixed reality) have been around since the early 1990s[77] with precursors since the 1960s. The technology has improved a lot since, and the current emerged technology offers much more convincing experiences than the earlier generations. But it suffers from some of the same

---

* Science fiction writer Neal Stephenson envisions the application of this technology in the future, where matter is generated in a somewhat similar way as 3D printers nowadays, in *The Diamond Age* from 1995. It may also have inspired some conspiracy theories relating to RFID chips circulating in our bodies through vaccinations, while in reality this technology is nowhere near that scale.

drawbacks (disconnect or mismatch with the real world,* motion sickness). Another current 'emerging technology' is AI, which is now in its third major incarnation.

In the original Internet of Things vision, as with Ubicomp, Pervasive, Sentient, Ambient, Disappearing, etc., IoT is about connecting all thinking things in the environment with the people in it. However, currently it is often understood in the watered down version of IoT as 'Things on the Internet', connecting a toaster or a fridge to the Internet has only very limited added value for people. Bruce Sterling wrote an insightful essay about these limitations, *The Epic Struggle of the Internet of Things*, in which he critically reflects on the role of the big companies that are trying to monopolise the Internet (through social media, search engines, applications) and their scenarios.[78] This is further reflected on in Chapter 12 *Understanding*.

To emphasise the importance of the interconnectedness, the distributed computing aspect, and most importantly the connection with the people, I started using the term **Electronic Ecology** or *e-cology* around 2003.[79] This was also in response to the many terms and misconceptions, though of course introducing another term might not be the most appropriate response, and certainly not a fancy *'e-'* word, but I found it useful. The notion of an ecology is a very important concept in biology, as it engages with the full complexity of all relationships and interactions between entities, animals, and their environment (discussed further in Chapter 7 and elsewhere). The key thing here is that potentially all nodes in this network can communicate with each other, even in a decentralised manner – data travels through the network without the need for a central server (the Internet is designed in this way too). This decentralised approach where all nodes in a network can communicate directly with each other, technically known as a 'mesh network', results in a very powerful and robust structure, and very similar to ecosystems in nature.

The terms ecology and ecosystems are often used in the context of interactive technologies; ecology is a very strong metaphor. The Ecological Approach to Perception as developed by psychologist J. J. Gibson since the 1960s emphasises this notion that people (and other animals) perceive their environment through their active engagement and interactions with it (discussed in depth in Chapter 9). The term electronic ecology similarly emphasises the interaction between people and their technological environment, rather than the interaction between electronic elements (embedded

---

* This is unless one has the equipment for motion and haptic feedback as envisioned in the Steven Spielberg movie *Ready Player One* from 2018, which also reflects on the dangers of becoming immersed in a virtual world, disconnecting with the real world. In another 2018 movie, *Johnny English Strikes Again*, Rowan Atkinson extends the notion of disconnect in a scene in where he accidentally leaves the simulation room and gets in some unique coincidental interactions outside and in places like a café and a bus, that align with the virtual images he thinks he engages with. It is more like Augmented Reality – to him at least, not for the people he unwittingly assaults.

digital computers). It is this broader and more pervasive concept of distributed and connected computer technology that *Understanding Interaction* is about. In Chapter 12 *Understanding* there is a further discussion of the electronic ecology and distributed interactive environments, and the role that the increasing 'datafication' (quantification) of everyday activities plays in this context.

INTERACTION AND EDUCATION

Since the early 2000s, several new bachelor and master degree programmes around design and interaction started to be developed at a range of universities in the Netherlands, and from 2003 I started teaching and researching in a number of these in part-time positions (while being based in Maastricht in the far south of the country, where I founded the MaasLab and was involved in several art and music projects).

The Industrial Design department at the Eindhoven University of Technology has a very radical programme, in response to (and partially driven by) the local consumer electronics and medical equipment company Philips, including some of the people I worked with there in the 1990s. The Philips company realised that they needed a 'different kind of engineer', people not just with technological skills and insights, or just design, but a combination of these, as well as the ability to do research into people's needs. The resulting programme was developed with a self-directed, competency-based learning model. I coached several student groups in projects (in some cases with Philips as a 'client') on interfaces for distributed interaction for the home, developed new modules for the master's programme, and for a while directed a multidisciplinary team of the Entertainment Unit, one of the four themes of the department, for education and research.

I also taught in several universities of applied science ('hogeschool', the more vocational training level of the two-tiered higher education system in the Netherlands, like the most of continental Europe), Mediatechnology and Communication and Multimedia Design (CMD) degrees in Amersfoort, Utrecht, and Arnhem. These are bachelor degrees, and recently these 'hogescholen' also offer master degrees and PhDs.

At the Academy for Digital Communication in Utrecht I wrote a small book *Interaction with our Electronic Environment – an e-cological approach to physical interface design* [2004]. The book was deliberately compact, in response to the increasing size of the HCI and ID textbooks.[80] This small book was based on the teaching I carried out before, and I used it in later teaching.

At the VU (Vrije Universiteit, Free University) in Amsterdam, working part-time as Assistant Professor ('universitair docent') from 2003–2007, I collaborated with Professor Gerrit van der Veer on developing and delivering a new version of the HCI teaching subject with a range of lectures, assignments, and assessment.[81] Here I also carried out PhD research on creating multimodal widgets (using visual, sonic, and haptic feedback, and measuring performance benefits of these interaction styles in a range of controlled studies), and developing a new framework for multimodal interaction design [2007] which is discussed after Chapter 6 *Communicating* at the end of Volume 1, and in further extended form in Chapter 12 *Understanding.*

My PhD thesis *Interactivation – towards an e-cology of people, our technological environment, and the arts* [2006] brought together all the strands of artistic, design, and scientific explorations of the previous twenty years. The PhD work was presented through the metaphor of a journey, following a meandering path along stepping stones (musical instrument design, architecture), reaching the foothills (multimodal HCI, haptic feedback, a physical interface design framework), and further explorations, not reaching a specific destination. I remember that one of the thesis supervisors seemed a bit disappointed with this, my insistence not to have ventured much further than the foothills – but it really felt like that. I was very aware (to continue the metaphor) that the real mountains were still only visible in the distance. And others invisible, obscured by clouds, or too far away.

The formal presentation of the PhD research and thesis in July 2006 at the VU was a public affair for which everyone dressed up, as is common in the Dutch system. The candidate is required to 'defend' the work in response to questions of a panel of inquisition (supported by two 'paranimfen', codefenders, a largely ceremonial but nevertheless crucial aspect of the local academic folklore). Experts from a wide range of backgrounds were invited on the examination panel, in order to ensure all disciplines explored in the thesis were represented; I am very grateful for the participation of these specialists, with backgrounds in computer science, HCI, music, and architecture (including Berry Eggen, Kas Oosterhuis, and Bill Verplank). I also created an exhibition, with all demos and interactive installations related to the thesis research, at the STEIM Studio for Electro-Instrumental Music in Amsterdam.

Soon after, I took up the position of Associate Professor at the School of Design at the Faculty of Design, Architecture and Building at the University of Technology Sydney (UTS).

At UTS I was given the opportunity to combine in one position much of the work I had previously pursued in an amalgamate of jobs at different places and levels and disciplines, including a freelance practice (and brought together in my PhD thesis). This position offered a solid foundation to explore many new areas, in design, theory, research, and the arts.

INTERACTION AND DESIGN

Craft disciplines traditionally work directly with the people who will use the product, which brings a focus on the interaction with the artefact.

Since the beginning of industrialisation (see Chapter 5) the notion of **design** became important – design as the structured approach to planning an outcome through a process rather than working directly and uniquely with the materials. Design is now an established discipline. A practitioner is a designer, design is an activity (design as a verb), and as a noun the design is the plan or blueprint or model that serves as a set of instructions for machines and/or craftspeople to carry out. (This is similar to the role of a score in musical practice, a way for a composer to communicate their ideas to a performer, separating the roles, and displacing the practice in time.) In everyday language nowadays, design often refers to aesthetic and conceptual qualities (and often, seemingly exclusively these qualities), perhaps best to describe that as **Design** (with a capital D) as opposed to **design** (with a lowercase d), the act of planning and modelling, with a focus on the people who will use the outcome of the design and manufacturing process. In other languages than English the different meanings of the word 'design' are referred to by separate verbs; design as planning is 'ontwerpen' in Dutch ('entwerfen' in German) and design as shape-giving is 'vormgeven' in Dutch ('gestalten' or 'formgeben' in German). The shape-giving of an object or system is not only about aesthetics, but also about usability, facilitating its utility through its form.

The design disciplines have traditionally been concerned with the interaction between people and the artefact, whether it is a building (architectural design), product (industrial design), clothing (fashion design), print (graphic design), or media design. This is the area in the middle at the top of the disciplines & interaction diagram in Figure 0.2.

Influential industrial designer Victor Papanek wrote in the late 1960s about the importance of usefulness and utility, and even sustainability and social relevance of design in his book *Design for the Real World* written in the late 1960s.[82] Papanek opens the book with a real eye-opener, by stating:

> There are professions more harmful than industrial design, but only a very few of them. And possibly only one profession is phonier. Advertising design, in persuading people to buy things they don't need, with money they don't have, in order to impress others who don't care, is probably the phoniest field in existence today.

(In our 'today', we would probably identify the invisible hand of social media companies as the 'phoniest' in the field.) Papanek states he was attacked by the design establishment at the time,[83] but the book had a lasting influence on many product designers. And although Papanek further writes "designers have become a dangerous breed", I must say that all industrial designers (particularly students) I have met in the last decades have been increasingly concerned about the environment and wellbeing of people using their designs; Papanek's books have contributed to that. Not enough though, it is shocking how badly designed many products in the real world are, still!

Interactivation and design research

In the first year at UTS I was generously provided with over 100 m² space in a prominent location (overlooking the central courtyard, near the café), a budget for renovation and construction of the physical infrastructure including lighting, and furniture (some custom made, mostly on wheels for easy rearranging), and awarded an internal grant to acquire state-of-the-art audiovisual equipment, tools, and materials. Extending from experiences of places I had worked before, including the interaction labs that I had set up in Amsterdam, Barcelona, and Maastricht, the Interactivation Studio was developed as a flexible and dynamic workspace suitable for research projects, teaching, workshops, studios, exhibitions, tests, presentations, and many other activities. This was facilitated by a flexible infrastructure and custom-designed building features, such as a smooth concrete floor throughout, cable trays in the floor, as well as a ceiling grid for cables and also suitable for hanging sensors and audiovisual equipment. The ceiling grid was even strong enough for people to hang on, as visiting PhD researcher Floyd Mueller (later professor, first at RMIT, and then at Monash University, in Melbourne) demonstrated by developing the first version of the *Hanging of a Bar* interactive installation here in 2009, a good example of what he calls an 'exertion interface'.[84] In the following decade the Interactivation Studio supported all these activities with an ever changing team of colleagues, students, interns, guest scholars, artists, designers, technologists, and visitors from collaborating companies and institutions, all interactivating away. A short article in the ACM *Interactions* Magazine in the series *A Day in the Lab* describes the space, and the activities and ideas behind it.[85]

In this period I developed several interactive installations, videoprojections, and performances around the campus and elsewhere (usually as part of festivals or curated exhibitions in galleries and museums). These were exploring various research questions, always aimed to investigate audience participation in interactive audiovisual installations. For instance, the range of *Facets* pieces were based on video projections of kaleidoscopic patterns, taking video footage I collected from dynamic textures in nature (deserts, weeds, cliffs, rivers, seashores, and underwater images), generated and manipulated in real time through physical sensors as shown in Figure 1.3. I developed the kaleidoscopic engine (in the graphical programming language MaxMSP/Jitter) as an algorithm with almost endless parameters to vary the image patters, to experiment with sensor inputs. A range of sensors were embedded in objects such as wooden blocks, balance plates, furniture, and in the environment, to explore new interaction styles in exhibitions.[86]

Overnight stays at Cockatoo Island, a former shipyard in the middle of Sydney Harbour, during the yearly events from 2011 to 2017 where the school took all first year design students for a three-day Design Camp, enabled me to try out interactive and performative video projections. These were sensor-driven stationary projections as well as moving projections,

Figure 1.3 Facets interactive projections of kaleidoscopic patterns through audience participations: Powerhouse Museum Play Late (2009), UTS Interactivation Studio interactive videowall (2008), Expanded Architecture Festival Carriage Works (2011), Design Camp Cockatoo Island (2014), Facets at Smart Light/ Vivid Festival (2009)

with the portable videowalker instrument that I developed over the years based on the earlier work in Barcelona and Maastricht. For these projections I was working together with colleagues and students, and we explored the rich history of the island. In addition to the industrial heritage, this included exhibitions of the Sydney Biennale, Outpost Street Art festival, and remnants or traces of film sets (such as *Wolverine* and *Unbroken*), as well as the ancient indigenous presence. The videowalker instrument is capable of real-time controlling the videos in size, shape, and speed through sensor controls mounted on top of the projector, which is connected to the laptop computer in my backpack (and an 8 kg battery, converter, and active speaker). Using an ever expanding bank of video clips of those previous events and traces, I was able to place these events back into the locations on the island, for a brief moment returning (a ghost of) the artworks and events to their place as shown in Figure 1.4.

In one instance the projection was layering accumulated recordings of a recurring event over time (spanning multiple years), presented in one

Figure 1.4 Videowalker portable projections performance, SEAM conference Sydney 2009

specific location. This dynamically resituating was also part of another project, *Traces – reading the environment*, which explored how traces of past events can reveal a history of previous behaviours and expressions of people in an environment.[87]

My main research project at UTS since 2009 has been on the development of interactivating exercises for rehabilitation purposes, particularly for stroke patients. Collaborating with many medical researchers, physiotherapists, occupational therapists, patients, engineers, and designers, from inside the design school, throughout UTS, and externally (hospitals, rehabilitation wards in Australia and the Netherlands, medical institutes, companies) and with internal and external funding, we developed a range of new devices, interventions, and approaches for interactivated rehabilitation practices. This included various versions of the Stepping Tiles, a modular system that gives feedback on a patient's balancing and stepping exercises, as shown in Figure 1.5.

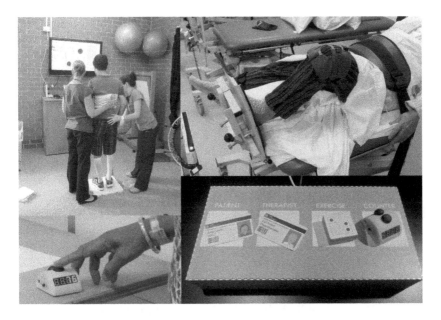

Figure 1.5 Some of the interactivated rehabilitation exercise modules (Stepping Tiles, knee brace, hand counter, and tangible exercise programming table)

The devices are often interconnected and distributed in an *e*-cology of sensors, displays, projections, and processors, in such a way that they are supporting patients, carers, therapists, and other stakeholders (see Figure 1.5). We realised that the key requirements of the interactive technologies were to allow for **customisation** (be able to create specialised interfaces for each individual patient, changing over time as their abilities shift), **independence** (to allow patients many opportunities to carry out their exercises, such as through tele-health support), and most of all, **motivation** (encouraging the patients to do the exercises, by making them more fun, game-like, and generally more engaging). All these developments were research based – including observations, interviews with stakeholders, and iterations of testing and redeveloping.[88]

An essential part of this project is the investigation of instant manufacturing techniques such as 3D printing, CNC milling, and laser cutting, as a means to create a paradigm of individual products to enable the optimal support of each individual patient and situation through customisation. This was the specific domain of expertise of my colleague Stefan Lie, and the topic of his PhD research.[89] The project is described in a range of publications in journals, book chapters, PhD theses, and presented at conferences and demos. PhD researcher Michelle Pickrell explored the interactivation of many stroke rehabilitation exercises, developing and testing a range of new wearable devices and feedback systems (short term and long term), working closely together with patients, physiotherapists, and other stakeholders.[90] The Stepping Tiles were part of a large-scale controlled study with over 400 patients, funded by the NHMRC (National Health and Medical Research Council in Australia). The study showed the positive benefits of interactivated rehabilitation tools.[91]

In separate collaborations across the school of design I worked with textile designer Cecilia Heffer on a range of exhibited interactive installations which combined Cecilia's delicate and intricate large-scale contemporary lace panels, and other patterned materials, with video projections, screens, and sensors. These included *InterLace*, a spatial interactive installation in the international Love Lace exhibition at the Sydney Powerhouse Museum from 2011 to 2013, and Pattern Stations in Craft ACT in Canberra in 2014 and at the TEI conference at Stanford University in 2015.[92]

Furthermore I collaborated with Deborah Szapiro and other colleagues and students from the Animation degree programme at UTS, for instance developing tools for interactivating animation which we publicly presented in a large-scale installation allowing audiences to manipulate animated sequences and motion parameters in real time through sensors embedded in everyday objects. I also collaborated on many occasions with the architect Michael Day who taught lighting design in the Interior Design programme and the MDes, on public light and video projection presentations (in the Vivid and Beams festivals in Sydney). Another project was a collaboration with researchers and designers from the Visual Communication degree

Figure 1.6 The Tangible Landscapes installation (2020)

(particularly Jacqueline Gothe) on interactive representations of indigenous knowledge structures, and especially the traditional bushfire practices and management (this is further discussed towards the end of Chapter 4).[93]

A recent installation, *Tangible Landscapes*, was presented in a gallery in Sydney for the TEI20 conference. It consisted of computers, screens, projectors, speakers, and interactive technology reflected on these experiences, combining elements and insights from the natural environment with the technological. Audiences could explore materials (found objects such as glass, pumice, burnt leaves, rocks, and driftwood) which were connected through sensors to the audiovisual content, as shown in Figure 1.6. This allowed participants to create reflective, meditative, and multisensory experiences.[94]

I mention these research and artistic projects (and the teaching, see below), as they formed the basis for the realisation that a broad exploration of knowledge in a range of scientific disciplines was needed, in order to create new foundations and knowledge by making connections across disciplines, and to create a broader picture of understanding interaction. Some of these projects (and others from this chapter) are discussed in more detail in other chapters in this book, when appropriate, to illustrate certain topics and approaches.

Interactivation teaching

An important part of my interactivation mission at UTS was to bring design for interaction into the teaching curricula across the faculty; the interaction topic was at that time almost entirely absent. I didn't think it was appropriate to start a new degree yet, first of all because I think it is a good habit when one is new to a country and a culture (with the experience of living, working, and studying in other countries before) to learn first, and as always I first wanted to get to know the situation and context well. Secondly, as discussed above, design for interaction is a crossdisciplinary subject that has relevance for all the different design fields (some examples have been presented earlier in this chapter): industrial design (interactive products), jewellery, fashion design (wearable interaction), graphic design, multimedia, animation, game design (interactive media), interior design (interactive spaces), architecture (interactive buildings), urban design (smart cities), and landscape design. Most of these disciplines were represented as undergraduate degrees in the faculty or covered through external collaborations. So the first phase was about bringing interaction into the existing degree programmes, through for instance specific subjects particularly in the industrial design degree and through guest lectures, workshops, studios, and electives in other degrees. I also connected to all the design degrees through the interdisciplinary design studies programme. This programme was quite unique for the UTS School of Design, some other design schools have a joint first year for instance, but this school endeavoured a connection across the design programmes, particularly in the first years of the degrees. This is where my colleague Kees Dorst (who started at the same time as me, as professor in design thinking; in fact he was the one who introduced me to the school) and I put our efforts. Teaching in this programme, it was interesting to notice how compartmentalised the degrees were, at least in the minds of the students; they seemed to have a very strict sense of what their design discipline was about, and even more particularly about what it was not. To support students' broader thinking, it is sometimes unavoidable and even desirable to be confronting, to encourage new learning. This might make students slightly uncomfortable, but at the same time creating a safe space to experiment and explore is essential.

I was also involved in setting up a new degree in Sound/Music/Design, across the design and humanities faculties, including co-organising an international symposium in 2008 before starting the degree as part of the development, and later hosting NIME, the international conference on New Interfaces for Musical Expression in 2010.

In 2011 we developed a new MDes (coursework masters in design) degree. Half of the degree was the core curriculum, the other half specialist studios throughout the three semesters. The Interactivation studio programme was a mixture of teaching and research, attracting students from a wide range of backgrounds, until 2017.

Together with my colleague Elise van den Hoven (now professor in the Faculty of Engineering and Information Technology at UTS), in the UTS School of Design we developed a 'sub-major' programme in Interaction Design. This sub-major was an option for design students to specialise in the 2nd and 3rd year of their bachelor programme, with a first year 'preamble' of a social design subject that I developed. Although the intention was to offer this to all design degrees, due to limitations in the curriculum structure in practice it was limited to product design students. It ran from 2014 to 2017. This Interaction Design subject followed the structure of the *Understanding Interaction* book, using shortened versions of each chapter (drafts), in weekly lectures, assignments, and tutorials for each topic. This helped developing the content of book, and I am grateful for the feedback and support of the highly motivated students in the subject. It also had a hands-on, technical component, in weekly workshop sessions for tinkering and theory.

Interactivation across design disciplines

As stated above, it makes sense to see design for interaction as a relevant area of knowledge and skills for all design disciplines, more than it being a design discipline on its own. All design, from the intimate scale and nature of jewellery and wearables to products and devices, to spaces, buildings, and further environments, all engage with people who eventually will interact with the technology, systems, and designs. Particularly in smaller design firms, companies, and studios, there is often no room to bring in the specific Interaction Design knowledge and skills. Ideally, therefore, many designers would have this knowledge (I am thinking again of all the regular struggles we have, with parking meters, microwave ovens, door handles, building layout, lighting, lift buttons, and almost any piece of software).

Interactivation and transdisciplinarity

The notion of Interactivation, reflecting on the broader relationship between people and technology, is applicable to a much wider range of disciplines, not just design: engineering, health, social science. Indeed we see topics related to design for interaction in different university departments or faculties, or in connecting entities. This works two ways; as discussed Interactivation draws on knowledge, methods, and insights from a range of scientific and practical disciplines, and it can then apply this through the interactivating lens, engaging with many applied disciplines and application domains.

This fits with the wider trend of connecting disciplines, in academic research as well as in society. It is increasingly urgent; there is a range of complex problems and situations in society that cannot be approached successfully from any discipline on its own. To explore this, UTS started with a radical new programme in 2014 of combining degrees – the Bachelor of Creative Intelligence and Innovation. This degree combines a range of existing degrees, with a new curriculum. More than a bunch of buzzwords, this new curriculum is aimed at engaging with the complexity of actual issues in society. And as a side effect, because it was seen as prestigious and certainly as hard work, requiring a large amount of dedication and engagement, it attracted some of the most ambitious and bright students. It works particularly well in the context of Australian universities, where many degree programmes have relatively short teaching periods (UTS has relatively short teaching 'semesters', effectively with classes for twelve weeks only, which gives students a lot of free time). Particularly the motivated students are often looking for bigger challenges. The programme fulfilled this need, and has been extremely popular with students. From the start I was involved in incidental lectures and tutorials, and from the beginning of 2018 on a full-time three-year secondment to what was by then the Faculty of Transdisciplinary Innovation (assuming the oxymoronic nature of this title was on purpose).

The structure of this double degree is that the students come from different core degree curricula, combined with the dedicated transdisciplinary second degree which explicitly aimed at connecting disciplines, practices, approaches, domains, etc. This model is often described as T-shaped: the rigorous specialist knowledge and skills of the stem of the letter T, and the generalist connecting knowledge and skills of the top bar, stretching out and linking to other disciplines. Ideally, and indeed in practice, it is more about tree-shaped rather than just T-shaped expertise, with multiple opportunities for connecting along the many branches of the trees reaching out (or rhizomes?). This is important, as discussed above; often the limitation of **multidisciplinarity** is that merely bringing different disciplines together in a team may not lead to cohesion. The term **crossdisciplinarity** is about a stronger connection, combining expertise from multiple disciplines. **Interdisciplinary** is about the overlap and connections, 'between disciplines', and specific domains that are created to link disciplines, ideally through multiple interconnections. The term **transdisciplinarity** is usually seen as the broadest, finding synergies and reciprocal relationships between disciplines. Taken literally of course the term would be more restricted, as 'trans' means across in Latin. In that sense transdisciplinarity would then be limited to merely passing through disciplines, similarly to transport (taking goods from one place to another), or the old Trans Europe Express (TEE) train network across the continent,* or the technical term transducer (turning one form of energy into another, like a sensor or an actuator), or transgender (transitioning from one gender to another, not a hermaphrodite). In practice, however, transdisciplinarity aims to connect in multiple and reciprocal ways – not only borrowing knowledge from multiple disciplines but synthesising and generating new knowledge. And by extension,

---

* Musically represented in Kraftwerk's famous album *Trans Europa Express* from 1977 (the iconic beat of the title track lives on in Pharrell Williams song *Marilyn Monroe* on his 2014 album *Girl*).

the importance of social impact, of developing and engaging with knowledge and insights that are relevant for society.

Understanding Interaction reflects this transdisciplinary approach (see the diagram in Figure 0.2). Interactivation is both a kaleidoscope and a prism, not just bringing together multiple disciplines in a clusterdiscipline (multidisciplinary, pluridisciplinary, polydisciplinary, or pandisciplinary, omnidisciplinary), or connecting between disciplines (interdisciplinary), or travelling through disciplines (crossdisciplinary), or plundering and looting from other disciplines along the way (plunderdisciplinary). The approach of *Understanding Interaction* is aiming to consolidate and synthesise new knowledge and approaches for developing and designing technology which is more suitable for people, society, and the environment. Otherwise, one might end up with a non-disciplinary approach, though a certain amount of discipline-agnosticism is necessary, we don't want to end up with a food fight – to continue the often used metaphor of a banquet, as for instance my former colleague Anne Cranny-Francis (professor in the humanities at UTS) used: attending the banquet, everyone can bring a dish without knowing what the others bring (multidisciplinary), or the situation where people will bring dishes for certain places in the meal, like starter or dessert (interdisciplinary), or the truly collaborative approach of communicating about all the food factors including ingredients, cuisine, etc. (transdisciplinary). Universities are usually very strongly structured in silos of independent and often isolated disciplines, and while this extreme specialisation has many advantages, it also has severe limitations, and often for anything new and worthwhile to happen we need to look into the connections between the disciplines for potential innovation. It is clear that to really understand interaction, the reductionist approach is not suitable. A more holistic approach is needed, just as can be seen in traditional cultures' knowledge systems (which is discussed at several points in this book).

It is not just the actual knowledge in a discipline that may be inaccessible to the outsider, but also the whole culture (attitudes, values, practices, and dogmas) of a discipline that may inhibit access or collaboration. This was already identified in the late 1950s, in a famous and often cited (and paraphrased) text *The Two Cultures* by C. P. Snow, in which he pitches science and humanities as mutually exclusive.[95] In practice, however, particularly in the later decades, individuals often connect between these two cultures – there are many examples of scientists who are also excellent musicians, painters, or writers. But of course, the point is still valid, that the sciences (and particularly their extension, the engineering disciplines) could be enriched by appreciation and engagement with the arts, music, and design, while it definitely could be of benefit for the humanities to be more aware and even actively embracing technology.

The influences between the cultures may be subtle, or strong, depending on the situation. A term often used for someone who spans different disciplines is the 'straddler' (interdisciplinary) or 'slash' (multidisciplinary).

For the mission of *Understanding Interaction* I often felt like a millipede straddler, and along the way there were many more hurdles and complications. One issue is that certain terms can mean different things in different disciplines; often a sign of isolation indeed, and in some cases even within a discipline, is that certain terms mean different things to different people. This is why in some cases it is appropriate and even crucial to introduce new terms with clear definitions, in order to support the process of 'straddling'. The Glossary section of this book (in Volume 2) aims to resolve this issue as much as possible, by covering the terms' multiple meanings related to different disciplines where appropriate.

The knowledge needed from a certain discipline to contribute to the knowledge of interactivation is often not just the basic textbook knowledge, as in 'Discipline X – Introduction 101 for Dummies', but often a careful selection needs to be made throughout the available knowledge in the field, not cherry-picking but creating a coherent contribution. For instance in psychology the ecological approach to perception (developed by J. J. Gibson in the 1960s and 1970s) is more applicable to interactivation than many elements of the more traditional cognitive psychology approach (Chapter 9 *Experiencing*, in Volume 2), or the recent insights in the importance of unconscious thinking and reasoning (not just about emotions and drives as thought previously[96], though still often about prejudices and misconceptions, as for instance the work of Daniel Kahneman and Amos Tversky has shown) (Chapter 10 *Thinking*). Another example is in the discipline of semiotics, where more recent developments in social semiotics and multimodality (as developed by Robert Hodge, Gunther Kress, and Theo van Leeuwen, among others) are most appropriate for *Understanding Interaction* (Chapter 6 *Communicating*, this Volume). And finally, the appreciation of 'pre-historical' cultures, technologies, and practices, which is explored in the next chapters on the history of humankind.

NOTES

1. *Understanding Media – the extensions of man* [McLuhan, 1964]. For a good overview see *Marshall McLuhan: the book of probes*, edited by his son Eric McLuhan, William Kuhns, and Mo Cohen [2003], with several contributions of W. Terrence Gordon, and designed by David Carson.

2. The term 'meme' was first proposed by evolutionary biologist Richard Dawkins in his book 1976 *The Selfish Gene*, as a shorthand for an idea. Internet versions of memes (often animated) are extremely popular, and like memes in literature and other cultural expressions evolve over time [Wiggins and Bowers 2015].

3. Some of these instruments are described in a chapter of my PhD thesis [2006, pp57–72] and my article in the *Leonardo Music Journal*, 'Electronic Musical Instruments: experiences of a new luthier' [2007]. See also the journal article by MIT colleague Joe Paradiso 'New Ways to Play:

electronic music interfaces' [1997], and the book by Eduardo Miranda and Marcello Wanderley, *New Digital Musical Instruments: control and interactions beyond the keyboard* [2006]. Jonathan Impett discusses the Meta-Trumpet and his algorithmic composition system in a conference paper [1994].

The technologies behind the instruments are discussed in my chapters for the e-book edited by Marcello Wanderley and Marc Battier at IRCAM in Paris, *Trends in Gestural Control of Music* [Bongers 2000], and performative aspects in an interview with the members of Sensorband (who used several of the instruments we collectively developed) in *Computer Music Journal* [Bongers 1998]. The first version of Michel Waisvisz's instrument *The Hands* was presented in a paper in 1985 at the *Computer Music Conference* [Waisvisz 1985], the versions I developed with him were based on these (the adjustable Hands for the Musikhochschule in Basel in 1987, and *The Hands II* in 1991 with which Waisvisz performed for nearly a decade), described in a recent article by Guiseppe Torre, Kristina Andersen, and Frank Baldé [2016]. These and other instruments we developed are discussed in an interview 'The Hand in the Web' in *Computer Music Journal* by Volker Krefeld [1990] and a chapter in Kevin Whitehead's *New Dutch Swing* about the avant-garde music scene in the Netherlands [1998] (thanks to Jos Mulder who gave this book to me). Michel Waisvisz passed away in 2008.

4. One book that discusses the notion of craft in digital design is by Malcolm McCullough, *Abstracting Craft – the practised digital hand* [1996].

5. In: *Handmade Electronic Music – the art of hardware hacking* by former STEIM co-director Nic Collins, now professor at The School of the Art Institute of Chicago, his own work and projects as presented in this book are an example of craft in digital design. 1st edition [2006, p192], 2nd edition, [2009, p251].

6. The development of haptic feedback technologies and experiments in applying this in musical instruments has been presented in several publications [Bongers 1994, 1998, 2000].

7. The elaborate design methods and the interaction styles we prototyped and tested with participants are reported in several publications [van de Sluis and Bongers et al. 1997], [van de Sluis et al. 2001], [Aarts and Diederiks 2006, pp63–65].

8. The Philips Force Feedback Trackball was a patented system [Engel et al. 1990], and the design and the research on haptic feedback has been presented in several publications by IPO researchers [Engel et al. 1994], [Keyson and Tang 1995], [Keyson 1996], [Keyson and van Stuivenberg 1997].

9. The four new multimodal interaction styles were 'company restricted', but Philips allowed one of the styles to be presented in a publication [Bongers et al. 1998].

10. Some good overviews of architectural projects using 3D modelling and interactive technologies (including the ones I was involved in) are:

Stephen Perrella (ed.), *Hypersurface Architecture II* [1999]

Peter Zellner, *Hybrid Space – new forms in digital architecture* [1999]

Kari Jormakka, *Flying Dutchmen: motion in architecture* [2002]

Lars Spuybroek wrote several books about his work, the best overview is *NOX: machining architecture* [2004]

Kas Oosterhuis has published several books about his ideas and projects in interactive, parametric architectural designs and actual buildings, particularly *Architecture Goes Wild* [2002], *Hyperbodies: toward an e-motive architecture* [2003], and *Toward a New Kind of Building – a designer's guide for nonstandard architecture* [2011].

11. The *Venice Architecture Biennale* exhibition was curated by Massimiliano Fuksas, under the theme of 'less aesthetics, more ethics', presenting 97 architectural offices of around the world. An extensive catalogue of the exhibition has been published, edited by Doriana Mandrelli [2000].

12. Curator architect Greg Lynn, who has been pioneering in non-standard architecture since the 1990s [Rappolt (ed.) 2008] interviewed the key participants, published as Apple books: *Muscle NSA, interview with Kas Oosterhuis* [Lynn 2014] and *H2O Expo, interview with Lars Spuybroek* [Lynn 2015].

13. The different versions of the Protospace design environment and laboratory, and the projects carried out by the Hyperbody group for about a decade, are presented in a book with contributions of all the participating architects and researchers [Oosterhuis et al. (eds.) 2012].

14. The *Soundnet* large-scale installation (10 m high and 6 m wide) on which the members of Sensorband (Atau Tanaka, Edwin van der Heide, and Zbigniew Karkowski) performed by climbing on to the structure, their motions picked up by custom-designed tension sensors, is discussed in the CMJ Interview (and shown on the cover) [Bongers 1998]. *Global String* was another room-scale instrument, consisting of two physical strings (10 m length) played in remote locations (V2 in Rotterdam and Ars Electronica in Linz by performers Atau Tanaka and Kasper Toeplitz, in November 2000), connected through the internet which became part of the instrument [Tanaka and Bongers 2001].

15. As defined by Étienne Grandjean from the ETH in Zürich, in *Fitting the Task to the Man: an ergonomic approach* [1980].

16. *Ergonomics: man and his working environment* [Murell 1965].

17. Stephen Pheasant, *Bodyspace* [1996 pp5–6].

18. *Fitting the Task to the Man – an ergonomic approach* [Grandjean 1980]. This reflects that it was still common to generalise all people to 'men' at that time. It is a bias which is a bit harder to avoid in English than in other languages; man (as in mankind, or human) sounds suspiciously un-PC but is effectively neutral, while in Dutch the word is 'mens' (or 'Mensch' in German) which can be either male (de man, der Mann) or female (de vrouw, die Frau). Grandjean himself by the way cannot be entirely blamed for the title of his book, as he wrote his famous book first in German in which the title was without any gender bias: *Physiologische Arbeitsgestaltung* ('physiological labour-design') in 1963.

19. The leading *Journal of Man-Machine Studies* was founded in 1969, later renamed to *Human-Machine Studies* but not until the late 1980s. It is currently known as the *Journal of Human–Computer Studies*.

20. My initial research and findings were published in a conference paper [Langdon et al. 2000], and later in more detail with further developments on Palpable Pixels at the *International Conference on Touch, Blindness and Neuroscience* in Madrid in October 2002, and a chapter in the book that was published after the conference [Bongers 2004] and also in my PhD thesis [2006, pp124–135].

21. My successor at the Cambridge University Engineering Design Centre, Faustina Hwang, led a range of excellent research projects around the use of haptic feedback for motion-impaired people [Hwang et al. 2003], and has since become professor at the University of Reading.

22. Metapolis Architects collective at that time consisted of Vicente Guallart (also involved in new media productions), Enric Ruiz-Geli (theatre productions), Manuel Gausa (publisher), Xavier Costa (writer), and Willy Müller. A book about the Media House project was published later [Guallart (ed.) 2005].

23. A book chapter 'Palpable Pixels: a method for the development of virtual textures' [Bongers 2004].

24. In a paper 'A Structured Instrument Design Approach: the Video-Organ' for the NIME conference [Bongers and Harris 2002].

25. The Design Space was based on the taxonomy developed at Philips and published in my chapter for the IRCAM *Trends in Gestural Control of Music* [Bongers 2000]. This Physical Interface Design Space (PIDS) is also presented in more detail in my PhD thesis [Bongers 2006, Chapter 6].

26. [Impett and Bongers 2001], [Harris and Bongers 2002]. See www.meta -orchestra.org.

27. Described in a conference paper 'Architecture and Motion: ideas on fluidity in sound, image and space' [Harris 2002].

28. [Moore 1965].

29. [Bush 1945], [Grudin 2012, ppxxx-xxxi], [Baecker and Buxton 1987a, p41]. This paper, as well as several other key papers mentioned in this section, are reprinted in *The New Media Reader* and contextualised with insightful introductory notes by the editors Noah Wardrip-Fruin and Nick Montfort [2003, pp35-47].

30. [Wardrip-Fruin and Montfort (eds.) 2003, p65], [Nørretranders 1998, p42]. James Gleick writes about these developments too in *The Information* [2011, p239].

31. [Gleick 2011, p238].

32. Licklider, *Man-Computer Symbiosis* [1960] is reprinted and discussed in *The New Media Reader* [Wardrip-Fruin and Montfort (eds.) 2003, p73], and other HCI handbooks [Grudin 2012, pxxxii], [Baecker and Buxton 1987a, p43].

33. [Gleick 2011, p245].

34. [Wardrip-Fruin and Montfort (eds.) 2003, p93], [Moggridge 2007, pp25–37], [Grudin 2012, pxxxiii], [Baecker and Buxton 1987a, p46].

35. The team published an extensive report about this research and development [English, Engelbart and Huddart 1965, pp91–92].

36. The video of the hundred-minute 'mother of all demos' is on YouTube, and on the web site of the Doug Engelbart Institute, www.dougengelbart.org, in full, annotated, and abridged versions, and the work is described in detail in a conference paper [Engelbart and English 1969].

37. [Wardrip-Fruin and Montfort (eds.) 2003, p109], [Sutherland 1963], [Baecker and Buxton 1987a, p44].

38. [Wardrip-Fruin and Montfort (eds.) 2003, p133], [Nelson 1965].

39. See for instance Ted Nelson's chapter 'The Right Way to think About Software Design' in *The Art of Human–Computer Interface Design* (edited by Brenda Laurel) [Nelson 1990].

40. [Bickerstaff 2014, p99]. A number of technologies and systems described in this section have been on display in the exhibition *Interface: People, Machines, Design* in the Powerhouse Museum in Sydney since 2014. The exhibition, curated by Campbell Bickerstaff, also compares the visual and industrial design style of Apple products since the late 1990s to the style of Braun consumer electronic products and other modernist designs of the 1960s. Apple's lead designer Jonathan Ive was an admirer of the influential designer Dieter Rams who worked for Braun, and Rams's minimalist style strongly influenced Apple's designs.

41. As described in the chapter by Alan Kay 'User Interface, a Personal View' [Kay 1990], in the book *The Art of Human–Computer Interface Design* [Laurel (ed.) 1990].

42. See the article of the Xerox Alto team in *Byte Magazine* [Smith et al. 1982].

43. Malcolm Gladwell recently wrote an article about these developments and reflecting on the nature of technology innovation [Gladwell 2011].

44. These developments are extensively documented in Bill Moggridge's book *Designing Interactions* through interviews with many of the key inventors and developers from that time [2007].

45. [Moggridge 2007, pp84–99].

46. *The User Interface – concepts and design* [Barfield 1993]. Lon and I taught a course together at the Interaction Design degree at the HKU.

47. *Developing User Interfaces Ensuring Usability through Product & Process* [Hix and Hartson, 1993].

48. [1995] (later editions of *About Face* have as sub-title *the essentials of interaction design* [2014]).

49. *The Humane Interface – new directions for designing interactive systems* [Raskin 2000].

50. *Interaction with our Electronic Environment – an e-cological approach to physical interface design* [Bongers 2004, p10].

51. In later printings since 1989 the title was changed to *The Design of Everyday Things*, and the new edition came out in 2013 with many additions [Norman 1988, 2013].

52. In: *Designing Interactions* [Moggridge 2007, p14].

53. In the first edition of the HCI book [Dix et al. 1993, pxiii], and the definition stayed the same in the second edition in 1998 and the third in 2005.

54. [Sharp, Rogers and Preece 2007, pv].

55. [Jacko (ed.) 2012, ppxxvii–lxi] (the introduction chapters have Roman numerals as page numbers).

56. [Jacko (ed.) ppxv–xvi].

57. [Cooper et al. 2014, pxix].

58. *The Inmates are Running the Asylum – why high-tech products drive us crazy and how to restore the sanity* [Cooper 1999].

59. [Sharp, Rogers and Preece 2002, pv].

60. [Kolko 2011, p15].

61. [Löwgren 2014].

62. [Carroll 2014].

63. [Fitzmaurice, Ishii and Buxton 1995].

64. [Ishii and Ulmer 1997]. See also *Tangible User Interfaces – past, present, and future directions* by Orit Shaer and Eva Hornecker [2012].

65. [Dourish 2001].
66. See the Proceedings of the NIME++ conference [Beilharz et al. (eds) 2010]
67. [Bongers 2002, 2003], [Bongers and Harris, 2002].
68. In a chapter in the book edited by Peter Denning, *The Invisible Future – the seamless integration of technology in everyday life* [2002, p252].
69. There are several articles on Sentient Computing [Hopper 2000], [Addlesee et al. 2001].
70. *The Invisible Computer: why good products can fail, the personal computer is so complex, and information appliances are the solution* [Norman 1998].
71. *The Invisible Future – the seamless integration of technology in everyday life* [Denning (ed) 2002]. Some of the authors are: Alan Kay who writes about that *The Computer Revolution Hasn't Happened Yet*, Bill Buxton who writes about *Less is More, (More or Less)* (see also his book, *Sketching User Experiences – getting the design right and the right design* [2007]), Ray Kurzweil about VR, Rodney Brooks about robots, people from Philips Research about Ambient Intelligence, and Bruce Sterling who writes about ubicomp which he introduces as: "Ubicomp is best understood as a very dumb and homely kind of digital utility, maybe something like air freshener or house paint. It's something the Joe Sixpack and Jane Winecooler can get down at their twenty-first century hardware store, cheap, easy, all they want, any color, by the quart (...) It is merely a common aspect of how normal people live". This is a great interpretation of Weiser's 'Calm Technology'. Sterling then supplements his essay about ubicomp with some outrageous and very clever scenarios, showing that as a Science Fiction writer he can go a bit further than the everyday scenarios we see so often in HCI (even in this *Invisible Futures* book).
72. There are several publications on Ambient Intelligence [Aarts, Harwig and Schuurmans, 2002], [Aarts et al. (eds.) 2003], [Aarts and Marzano (eds.) 2003].
73. [van de Sluis and Bongers et al. 1997].
74. [Eggen (ed.) 2002], [de Ruyter (ed.) 2003].
75. [Aarts and Diederiks 2006].
76. Gershenfeld and colleagues explained their vision in an article 'The Internet of Things' in the Scientific American [2004].
77. See a selection of the many books of that time about Virtual Reality [Rheingold 1991], [Aukstakalnis and Blatner 1992], [Kalawsky 1993]. One of the first applications of a glove-like input device was presented in the late 1980s [Zimmerman and Lanier et al. 1987].
78. The essay 'The Epic Struggle' actually opens with: "The first thing to understand about the 'Internet of Things' is that it's not about Things on the Internet. It's a code that powerful stakeholders have settled on for their own purposes." And in the next paragraph "It disguises the epic struggle over power, money and influence that is about to ensue." [Sterling 2014].
79. In a small book that I wrote at the time, *Interaction with our Electronic Environment – an e-cological approach to physical interface design* [2004], in a book series called 'Cahier' of the communication and journalism school at the Academy for Digital Communication where I worked at the time – a 'cahier' is a kind of notebook in the Dutch language, from a French word.
80. This is both in volume and weight, for instance, the first edition of the HCI book by Alan Dix et al. was about 750 grams (570 pages) [1993], and as mentioned I was able to read the whole book in the mid-1990s, but the second

edition weighs about 1500 grams (638 pages) [1998] and the third 1800 grams (834 pages) [2004]. The ID book by Helen Sharp, Yvonne Rogers, and Jenny Preece more than doubled from a modest 800 g (519 pages) in the first edition [2002] to 1700 g (773 pages) in the second edition [2007]. The Handbook of HCI, indeed always meant to be more of an encyclopaedic nature, in its third edition weighs about 3.5 kg (I needed to upgrade my scales to establish this), 1452 pages [Jacko 2012]. Smaller and more manageable (and possibly more readable) books were often either biased and selective, or lacking in depth and rigour.

My modest booklet *Interaction with our Electronic Environment – an ecological approach to physical interface design* [2004] was an attempt to find a balance between rigour and selectiveness by focussing on what I thought were some of the most important issues (all chapters started with an 'I'): Introduction, Interaction, Interface, Information, and Integration. The result was only fifty pages (157 g!) and a video-CD with image-essays (or video illustrations and illuminations) of reflections on interactions with technological environments.

81. We published a book chapter about the course [Bongers and van der Veer 2009].

82. *Design for the Real World – made to measure* published in 1971 [Papanek 1971].

83. After the first edition of Papanek's book [1971], he reflected on its impact in the preface of the second edition of *Design for the Real World* in 1984, which has the subtitle *Human Ecology and Social Change.*

84. See the web site of the group, www.exertiongameslab.org for an overview of their very physical interaction projects. The *Hanging of a Bar* installation was further developed by Florian and his team, and presented in several shows and as a CHI paper [Mueller et al. 2012]. Florian also worked with us on his framework for exertion games, which was published as a CHI paper [Mueller et al. 2011].

85. ACM *Interactions* magazine in the series 'A Day in the Lab', Jan/Feb 2014.

86. An overview of these interactive kaleidoscopic video projection projects can be found in a journal paper [Bongers 2012], and a publication focussing on the audience interaction studies [Bongers and Mery-Keitel, 2011] of PhD researcher Alejandra Mery-Keitel [2013].

87. Some of these works are presented and discussed in publications [Bongers 2011, 2012, 2013], and in future articles.

88. The interactivated rehabilitation work has been presented at several conferences, such as the *TEI* [Bongers et al. 2014] and the *Conference on Pervasive Computing Technologies for Healthcare* [Donker et al. 2015], and two book chapters [Bongers and Smith 2011], [Bongers et al. 2014].

89. Stefan's PhD thesis, which I supervised, is titled: *Assisting Product Designers with Balancing Strength and Surface Texture of Handheld Products Made from 3D Printed Polymers* [Lie 2020].

90. Michelle's research was presented at several conferences [Pickrell et al. 2015, 2016, 2017], and in her PhD thesis [2020].

91. This work was carried out by a range of researchers from different universities and hospitals, including one in Adelaide, led by Professor Cathy Sherrington from the University of Sydney and published in several journal papers [Hasset et al. 2016, 2020].

92. The *InterLace* interactive video projections and lace material was chosen for the *Love Lace* exhibition of the finalists for the International Lace Award at the Powerhouse museum in Sydney, curated by Lindy Ward, from July 2011 to October 2013, received a Highly Commended award [Ward 2011]. A paper about the Pattern Stations was published in the proceedings of the *TEI15* conference [Bongers and Heffer 2015].

93. Our main collaborator in this work is Victor Steffensen, active advocate of appreciating indigenous knowledge, and who recently published a book to tell the story from his own aboriginal background, *Fire Country* [2020].

94. A short paper discussing the conceptual background of the *Tangible Landscape* pieces in this installation was published in the proceedings of the *TEI20* conference [Bongers 2020], as part of the Arts Track. The installation was presented in the Tin Sheds Gallery at the conference venue of the University of Sydney (the other venues were at UTS), 10–12 February 2020, in the exhibition curated by Karen Cochrane, Thecla Schiphorst, and Deborah Turnbull Tillman. A video documenting the installation and audience interactions and responses can be seen on my Vimeo channel (www.vimeo.com/bertbon/videos).

95. *Two Cultures* was delivered as a lecture by C. P. Snow at the University of Cambridge in 1959, extended from an earlier publication in 1956. A second edition was published in 1964, with an added section in which the author reflects on the impact of the lecture, as an expanded version of *Two Cultures and the Scientific Revolution* [Snow 1964].

96. See for instance neurologist Joseph LeDoux *The Emotional Brain – the mysterious underpinnings of emotional life* [1995] or the view from cognitive psychology in *Emotional Design – why we love (or hate) everyday things*, by Donald Norman [2004].

Chapter 2

Evolving

Technocultural periods and technology categories

To understand interaction between people and technology, it is essential to look at the history (including the 'prehistory'), as it is both fascinating as well as informative to investigate our origins. For millions of years, we have been shaping objects and environments to suit our needs and to further develop our mental and physical abilities. There are strong reciprocal relationships between the development of culture, technology, biology, behaviour, and environment, and this needs to be explored in order to understand interaction. In this context of understanding interaction, the aspects of culture I am focusing on in the historical perspective are the design and use of tools and artefacts, and the development and use of language (oral and literal). This starts in prehistoric times, several million years ago.

Prehistory means the history before the invention of writing. There was certainly history in the prehistory, and it is essential for understanding interaction to look at the culture, tools, and developments from the first use of artefacts as a way of externalising the thoughts and desires of the early human and hominin species. Although it is much easier to study 'history' because of the availability of the written sources, there is great richness and vast knowledge in the preceding oral cultures. In fact, as we will see, writing did not replace oral culture, oral culture existed (and still exists) in addition to written culture, and it is worth studying it in order to attempt to get more complete insights.

EVOLUTION AND TECHNOCULTURAL CHANGE

When Charles Darwin and Alfred Russel Wallace first presented their ideas on (biological) evolution in the mid-19th century, people found it hard to believe this idea of the evolving of species, because until then it was assumed that species were fixed. And indeed it is hard to see species change, in the short time spans of our direct experience. But with a bit of effort, practice, and insight, it is possible to acknowledge and understand this slow pace of biological evolution.

Evolution is the development over time of an organism in an ecosystem – the interactions between each other and all the other species directly related

DOI: 10.1201/9781315373386-2

to them, and their habitat. Gradual changes take place based on what Darwin called natural selection: under *variation* (often due to random changes in genes) certain traits can be developed or extinguished. This process can be influenced and accelerated by humans through artificial means, as used in agriculture and the domestication and appropriation (or misappropriation) of animals. This is how Darwin studied it actually, in his garden at his estate in England, in much more detail than what he observed during his now famous journey with the Beagle around the world early in the 1830s.[1]

One reason for my focus on the 'pre-historical' developments is that a lot can be learned from the co-evolving of biological and technocultural changes. Some of the development would have been natural (a change in habitat – the primates that were our ancestors coming down from the trees to start living in the open savannah), but other changes were related to the use of technology (the shape and function of the early humans' hands as a response to tool use). Furthermore, there is a strong relationship between technology and culture. Technological inventions lead to behaviour change and variation in cultural practices.

Biological evolution is relatively slow. It often takes many generations (hundreds of years in the case of humans) for a change to occur, although for the keen observer (like Darwin and Wallace in their time) the evidence of the changes can be seen in certain species. Cultural and social change is much faster – these changes can be easily observed within the span of a life-time. Technological development can be even faster. There is a strong inter-relationship between technology, culture, and society. This can be called technosociocultural development, an evolution of gradual development. Because evolution is an accumulative process of inventing and discovering (by natural or artificial means), it speeds up as new inventions are based on the accumulation of earlier developments. This is potentially dangerous, like the proverbial runaway train. In order to predict where the developments are going, it is essential to study where they come from. But due to the exponential increase in the speed of developments (due to the accumulation of inventions) it is still difficult to predict the future.

Presented in the historical overview in this and the next chapters is the development of biological and technocultural changes. The first millions of years are particularly interesting. Though the development is slow compared to later developments, and particularly the amount of changes happening in our contemporary society, it is a process of profound co-evolution that we need to engage with in order to understand our current situation better. In contrast to the pace of the biological evolution, the technocultural changes are very rapid and radical.

TECHNOCULTURAL PERIODS

For the purpose of understanding how we interact with technology, it is useful to discern more stages than the traditional distinction between Stone

Age, Bronze Age, and Iron Age. This classification from the early 19th century is a good illustration of the exponential nature of development (the Stone Age is about 700 times longer than the Iron Age), and it is also combined with the traditional bias of that time of ignoring earlier developments and overemphasising the contemporary. Furthermore, it is worth mentioning that in the 'Stone Age' also wood, bone, clay, and other animal and plant materials were used.

The **tool period** of the prehistoric interaction (manipulation and communication), starts about 3.5 mya (million years ago), and is covered in Chapter 3. In the tool period we can distinguish the four main phases: tool or artefact *using* (opportunistic use of found or modified objects), *making* (the dedicated making of objects for a certain task), *designing* (objects made with a planned and projected use in mind), and *composing* (the design and making of composite objects). Tools and language developed concurrently for millions of years, in a complex interrelationship.

People lived mostly nomadically, primarily as hunter-gatherers. This is the Pleistocene Epoch of the quaternary geological period.

During the **settlement period**, discussed in Chapter 4, with the start of agriculture about 12,000 years ago, and the establishment of villages and towns, composite tools further developed in *mechanical* structures. These mechanical structures were first *passive* (driven by human action) and then *active* (driven by an external power source such as another human being, animals, water, wind). This is the Holocene Epoch of the quaternary geological period.

The **industrial period** started around 500 years ago and is discussed in Chapter 5. During this industrialisation, mechanical systems were integrated with new power sources such as steam engines, and later the internal combustion engine, and eventually electric motors. These *active mechanical systems* are machines. The distinction between passive and active is relevant because the interaction with a passive system (where the human is the operator as well as the driver through the provision of energy) is very different from an active system (a machine, where the kinetic energy is not supplied by the human, but externally sourced). As an illustration of this difference one can think of the difference in interaction between a bicycle and a motorbike, or a pedal sewing machine versus an electric sewing machine.

I am following commonly accepted key periods of change, which seem the most relevant for understanding the interaction between people and technology. The historian Yuval Noah Harari in his book *Sapiens, a brief history of humankind* [2014][2] uses the labels Cognitive Revolution (70,000 years ago), Agricultural Revolution (12,000 years ago), and Scientific Revolution (500 years ago). These roughly correspond with my classification, but I am more focused on the key influences on the relationship (interaction) between people, culture, and technology which started much earlier, as mentioned. It is also important to bear in mind that what we often refer to as historical 'revolutions' were often much more gradual and slow changes, cumulative and emergent, usually based on the interdependency of a large number of developments and factors, and stretching over a long time span.

For understanding interaction in a historical perspective, it is important to look at this larger time frame. As will be discussed below, several millions of years of the different stages in interaction with tools, and developing communication skills between people, were a concurrent process and it is plausible that language emerged as a feature of this process. Harari assumes that a genetic change occurred about 70,000 years ago which lead to the 'cognitive revolution'. This is also the view of the psycholinguist Steven Pinker who defends the position that humans uniquely possess a 'language gene' and a 'language instinct' (the title of his book from 1994).[3] The prominent linguist Noam Chomsky since the 1950s has argued that there is a 'universal grammar',[4] thought to be a feature of the neurological structures in the human brain, to explain the indeed striking similarities between almost all human languages.

There is an ongoing debate about this 'language gene'. Linguist Guy Deutscher gives a short overview of it in his book *The Unfolding of Language – the evolution of mankind's greatest invention* [2005]. The contradiction in the title of Deutscher's book is on purpose; "language is mankind's greatest invention – except, of course, it was never invented."[5] No evidence for such a genetic change has been found, and this explanation is a typical linear and causal way of thinking about things. As a species, we (as well as our ancestors) have been communicating in some form or another for millions of years, through images and increasingly abstract oral language for at least some hundreds of thousands of years, in writing for a couple of thousand years, and through the use of multiple media in the last hundred years – over increasing distances, spanning the whole world, even reaching into outer space. Rather than this being an invention, or as a result of any clearly identifiable causes, it seems perfectly possible for languages to *emerge* from the increasing complexity, accumulating developments and evolutions, in response to selective pressures and needs (social, practical, and environmental).

A particularly salient feature of our language skills is the ability for recursion, or subordination as it is sometimes called, which is the technique of inserting statements or clauses within the sentence structure, as a way of near-simultaneously presenting a number of ideas or aspects of an idea, in nested structures (this sentence contains several examples of recursion).[6] It is also a common technique in computer programming, when a function definition can refer to the function itself.[7] A common visual example is where a copy of an image is embedded in an image itself (with another copy of the image embedded, leading to infinite regress) such as in the old Droste cocoa tin (we had one at home in Holland). It is the kind of feedback loop which is common in audio systems, where a microphone picks up its own sound from the speaker, and a technique which many electric guitar players use with great skill.

Recently this nativist or innatist (language being innate, not just learned) notion is being refuted by linguist and anthropologist Daniel Everett [2012]. Everett is an expert in the language of a people in the Amazon, the Pirahã (pronounced as pee-da-han),[8] who, among a number of fascinating aspects of their language (which will be discussed later, in Chapter 6 *Communicating*, and in Chapter 11 *Expressing*), uniquely do not possess this ability for recursion.

Furthermore, it seems plausible that our evolution of abilities to manipulate objects, use tools, and make and even design artefacts (a process of a few million years) can have supported the development of our language abilities.[9] Simply put – the object manipulated as the noun, the action performed as the verb, and the order of actions and manipulations as a form of grammar. Take for instance the everyday activity of getting dressed, putting on your underwear before putting on your trousers, putting on your socks before putting on your shoes – these activities need to be carried out in a certain order. Other examples would be the procedures involved in cooking a complex dish, the activities related to decorating a house, etc. It is easily imagined how recursion can occur in these examples.

In these chapters, therefore, I am covering the development of tools and language, or technology and culture, in the simultaneous way as it has evolved and developed.

CLASSIFICATION OF TECHNOLOGY FOR INTERACTION

As a guiding structure for studying the history of technological developments in relation to artefact design and culture, I am using a classification of technologies which I developed over the years and specifically as applied to musical instrument design (which was the field that first inspired the core of this classification).[10] I am focusing on the aspects of technology that are relevant for human development of tools and the co-evolution of language and culture, which is important for understanding interaction. The categories are described in more detail later, in Chapter 5, including the sub-categories and examples in the section *Technological Developments*. The emphasis is on the human interface for each of the particular technologies.

The technologies in these categories have not been invented and developed in a linear history, in each category we can distinguish key inventions in different time periods, and many interrelated developments.

Mechanical technology

The oldest technology in many ways is **mechanical technology**. The first technological artefacts were **objects,** and in order of development over time, either **found, planned, designed,** and eventually **composed,** in their appropriation as tools. The oldest artefacts were made by human ancestors millions of years ago. Later technologies in this category are mechanical contraptions that use levers and cogs for transmission and conversion of forces applied in machines (particularly after the invention of metallurgy), first **passive mechanical machines** (driven by the person operating it) then **active mechanical machines** (driven by an external power source such as water, wind, and animals (in the settlement period), and later by engines in the industrial period).

Chemical technology

There are many examples of **chemical technology** we interact with; historical examples are medicine, food, and fire which have been in use for hundreds of thousands if not millions of years. In the industrial period steam power, the combustion engine, and photography (and later film) are all examples of chemical technologies.

Electrical technology

Although the existence of electricity in nature was known for thousands of years (examples are lighting and static electricity), the use of electricity in the tool period and settlement period is very limited.

In the industrial period **electrical systems** such as the telegraph and light bulbs were introduced, and later **electronic systems** (which can modulate and transform electrical signals, such as radio, amplifiers, and synthesisers) – first **analogue electronic** and later **digital electronic**, and eventually leading to the **computer**.

Optical technology

Early uses of **optical technology** were through glass-like materials which occur in nature. These have been used in tool making since prehistoric times, such as obsidian obtained from volcanic sites or meteorite impact craters.* The Aboriginal people of Tasmania were using glass 27,000 years ago, over 100 kilometres from its source.[11] Later, people were able to make glass artificially, during the settlement period, and lenses were produced with increased sophistication. In the industrial period lenses were further developed, subsequently applied in the invention of microscopes and telescopes (and later binoculars), and later in cameras (in combination with *chemical technology*, see above). In current applications optical technology is important in the use of glass fibre-optic cables, which facilitate all high-speed and high-bandwidth communication of *digital electronic technology*.

---

* Obsidian is also a sought-after material in *Minecraft* to create a nether portal, apparently.

Magnetic technology

Magnetic materials have been known since the settlement period (they occur in nature), and later applied in tools such as the compass, which detects the earth's magnetic field, pointing within a few degrees to the North Pole (first applied in late mediaeval times).

Later developments used **magnetic technology** in combination with *electrical technology* to create electromagnetic systems used in motors and generators, as discussed below.

Nuclear technology

Nuclear energy has been used in power plants since the mid-20th century. **Nuclear technology** has also been applied since then in atomic bombs and other nuclear weaponry, inventions with intentional (one hopes) non-use, other than a deterrent. Since its discovery at the end of the 19th century, it has been used in X-ray systems for medical analytical purposes, and in systems for determining the age of objects or places (based on the known pace of transition of electron distribution of specific atoms), and other analytic tools.

TECHNOLOGICAL FUNCTIONS

In addition to this categorisation of technologies by type, are the functions of technology. The first distinction I make is between technology that uses energy transduction for **communication** (of information, through signals or data), and for **power** (the work that it does). For each, there is **generation** (or conversion rather, as the total amount of energy is always constant in a closed system – according to the law of conservation of energy), **transmission, reception**, and **storage**. To give an example with electrical technology, communication can be facilitated with the telephone, where a message is *generated, transmitted*, and *received*. The electrical information can be *stored* on disks or tapes (analogue electronic technology, the now more common digital communication, is discussed in Chapter 5). On the other hand, energy transduction of power would be the *generation* of electrical energy in a power plant (where one form of energy, for instance, heat or kinetic energy, is converted into electrical energy), *transmitted* over power lines, and *received* in the home to power electrical appliances. Electrical energy can also be stored, in a battery.

Of course electrical communication (all communication for that matter) uses energy. Technically, they may seem the same thing, but functionally this distinction is important: energy in the form of information, and energy that drives things, that gives power (and of course, as any electrical engineer would know, it is even possible to code information in energy in such a way that it can also power things, for instance sending digital signals through the electrical power wires in the home). This is discussed in more detail later in Chapter 5 – *Industrialising*.

TECHNOLOGICAL CATEGORIES AND INTERACTION

It is important to distinguish these categories for understanding interaction, because each has its own inherent way of being controlled, interacted with, by people. In other words, the underlying technology determines aspects of the interface. The way an active mechanical system is interacted with reflects the technology that drives it, for example, a steam train interface is different from the way a diesel train is controlled, and yet again different in the case of the electric train. A clear example is a rotating dial, the mapping of which is dependent on the underlying technology. In mechanical technology, a dial is based on the screw thread, which means it opens when turned anti-clockwise. In electrical technology, the convention for open (mapping of rotational movement to increasing value of the parameter controlled) is clockwise. This is effectively an arbitrary convention, as the electrical dial is based on a potmeter, which could be wired in any way to create either mapping. In mechanical technology, one could say that the direction of the screw thread is also based on convention and therefore arbitrary – I have always wondered where it came from. I have noticed that in nature all thread like structures seem to have the same direction (going up to the right), in shells, climbing plants, and trees. The exceptions in mechanical technology usually have a practical reason, for instance on the pedals of a pushbike, the left pedal has the opposite thread to normal; otherwise it wouldn't stay in. This is very confusing when encountered for the first time – but it is still arbitrary, there is nothing 'natural' about normal thread direction. It is a learned thing, children have to experiment to find the right direction, and so must laypeople. On gas bottles the thread is also reversed, presumably to avoid tampering, so it is an example of an interactional aspect. However it is one that doesn't make sense, because it is a learned thing, so it wouldn't stop a child from accidentally loosening the gas hose.

The technological categories are often combined in **compound technological systems**, for instance, the combination of magnets with electrical coils leads to electromagnetic technology motors and generators, and the addition of engines (steam, combustion, or electromagnetic) to mechanical systems leads to the category of active mechanical systems, and other examples given above.

Most current technological systems are a combination of several technologies. For instance, a car is primarily a mechanical structure, but the internal combustion engine is based on a chemical principle (burning, or rather, exploding the fuel), literally sparked by electricity, with added electrical systems (headlights, indicators), electronic systems (audio system), digital electronic (controlling the fuel injection, and other functions), and computer systems. A traditional photo camera also has a central chemical process (the film that gets exposed to light patterns capturing the image), optical technology (lens), and mechanical technology (shutter, adjustments). Later cameras included electrical (flash bulb), electronic (light measuring device), and eventually digital electronic subsystems. Currently cameras are digital, based on an image-capturing sensor (an analogue electronic component, read digitally), but still in combination with optical and mechanical technologies.

All these technologies have their own inherent ways of interacting, which in some cases lead to a potentially confusing amalgamate of interaction styles. I have often used the car as an example in my publications and lectures; each newer subsystem that is being 'tagged on' has its own interface and interaction style, and it only becomes integrated over time. The addition of the radio to the car is a good example. This analogue electronic technology took decades to integrate with the rest of the car controls (nowadays one has dials and buttons for controlling audio volume and station or track selection on the steering wheel). Similarly, early navigation systems were tagged on and again contributed to driver distraction, but they are now more integrated, as is the phone through its Bluetooth connection to the car's sound system. A holistic design approach would benefit the interaction. Of course, it is not just for technological and historical reasons, but vested interests of diverse industries that oppose a unified design approach, which will be further discussed later.

Before we can get into the car though, we need to go back in time and study the first use of tools, millions of years ago. In the next chapter, it is discussed how it all started.

NOTES

1. Geneticist Steve Jones (Professor at UCL in London) wrote an article with an excerpt in the Guardian from his book *Darwin's Island – the Galapagos in the garden of England* [Jones 2009] entitled 'We Ought to be Exterminated', published on 19 January 2009. The article and the book draw attention to the importance of Darwin's research on his estate of Down House, after returning from his journey. Jones contrasts this with the stereotypical view including that he was reluctant to share his view in order to avoid trouble with his contemporaries and particularly the church [Bregman 2013, p202] – while in fact his essential research in the decades after the journey was carried out to prove the theory.
2. Historian Yuval Noah Harari (Professor at the Hebrew University in Jerusalem) wrote the book first in Hebrew based on his lectures over many years, published in 2011. Harari later translated it from Hebrew into English (with the help of John Purcell and Haim Watzman), published in 2014.
3. Steven Pinker, *The Language Instinct – the new science of language and mind* [1994].

4. See for instance the textbook by Vivian Cook, *Chomsky's Universal Grammar – an introduction*. [1988].

5. Guy Deutscher, *The Unfolding of Language – the evolution of mankind's greatest invention* [2005, pp16–18, p310]. The deliberate contradiction in the subtitle is on the first page of the book.

6. [Deutscher 2005, p254].

7. The book *The C Programming Language* by Brian Kernigan and Dennis Ritchie, in the topic index under 'recursion' lists not only the page numbers where the topic of recursion is covered but also the page of the index where the term occurs (page 269, in the 2nd edition, of 1988).

8. The pronunciation of the name Pirahã is mentioned in an extensive article about Everett and his research, including interviews with other linguists, in the New Yorker [Colapinto, 2007].

   Dan Everett's book about this topic is *Language – the cultural tool* [2012]; the discussion on recursion is on p288.

9. Frank Wilson writes about this in *The Hand – how it shapes the brain, language, and human culture* [1998], and see also *Tools, Language and Cognition in Human Evolution* [Gibson & Ingold (eds.) 1993], further discussed in the next chapter.

10. The categorisation of technologies is presented in earlier writings on technology [Bongers 2004, 2006 Ch. 1.2] and musical instruments [Bongers 2007].

11. The use of obsidian is discussed by Scott Cane in *First Footprints – the epic story of the first Australians*. [2013, p124/p270n40].

Chapter 3

Creating

Dedicated tool making and the emergence of language

The relationships between biological evolution, tool development, and the emergence of language are explored in this chapter.

I have chosen not to start with the beginning of time, the big bang, 13.8 billion years ago. Nor did I want to start with the forming of planet Earth, around 4.5 billion years ago.[1] Or the start of organic life, about 3.8 billion years ago, first as single-celled organisms (prokaryotes such as bacteria) which only after another 2 billion years (around 1.5 billion years ago) developed in multicellular organisms (eukaryotes). From there further multicellular organisms started to develop, particularly around 550 mya (million years ago), the so-called Cambrian explosion.[2]

We also had to wait for the dinosaurs to become extinct, about 66 mya, due to the impact of an asteroid leading to a long-lasting change of climate unsuitable for most reptiles. Until then the earth's terrestrial environment was for millions of years dominated by the larger reptiles, eclipsing any possibility for mammals to develop. Only after the demise of the dinosaurs, mammals had a chance to proliferate, leading to the primate families that the humans descend from.

So, let's start at the beginning of human development, from the first hominins. I am using the term **hominin** to refer to all human-like species after the split from other apes (about six mya), as is now common in literature, but the terms in taxonomies change over time. The term 'hominids' (Hominidae) includes the great apes as well, but sometimes the terms are used as synonyms, depending on the discipline, particularly in older literature. Currently, the term hominids covers all great apes (chimpanzees, bonobos, gorillas, and orangutans) and hominins. Hominins include the genii of *Homo* and *Australopithecus* (all of which are extinct, except for *H. sapiens*).

ARCHAEOLOGY AND PALAEOANTHROPOLOGY

Archaeology is the study of human cultures in the past,[3] or more specifically, the study of the human past through the material traces of it that

DOI: 10.1201/9781315373386-3

have survived.[4] Due to the inevitable incompleteness of the fossil record and other remains of early humans, the knowledge of our past is patchy and full of (however educated) speculation, and there is a large variety in the interpretations of the finds. Dating of the finds is often compromised and based on contextual factors, though the dating techniques have improved over time (which means that dates in older literature often need to be adjusted). Genetic analysis is another technique that is used. The field of archaeology uses very sophisticated methods, and has impressive knowledge and techniques built up over the years. As **anthropology** studies human societies and cultures, **palaeoanthropology** studies all the (early) humans and hominids from the past.

Consensus about many topics in archaeology and palaeoanthropology changes over time, often due to new finds and insights gained (including improvement of dating technologies). Most of the topics are not so relevant for our case, so I am trying to focus on the archaeological and palaeoanthropological knowledge that is needed to understand the history of interaction.

There is a complex network of evolutionary interrelationships between the factors of bipedalism, locomotion, gestation period, brain size, cognition, metabolism, diet, mastering of fire, development of language, tool making and using, transport, senses, amount of body hair, making of clothes, etc. In the following sections this will be discussed in detail. What can be clearly seen is a co-evolution between behaviours and biological adaptations. Take for example the relationship between gestation period and locomotion posture. In order to be able to consistently walk upright (bipedalism, walking on two legs), the human pelvic bone structure and its shape had to evolve to support this way of locomotion. This means that the birth canal (the space for the child to be delivered) is compromised, so it has to be born before it is fully developed. Compare this to four-legged animals such as cows or horses, where the newly born calf or foul is ready to stand on its legs straight after birth. An extreme case of the effect of bipedalism can be seen with kangaroos, whose habit of skipping has decreased the pelvic space even further, and the little 'joey' that comes out is effectively a foetus that has to grow and develop further in the mother's pouch. This is a good example of an interrelationship, between the evolution of the pouch and the evolution of locomotion technique based on skipping.

Of further influence on the human gestation times is our disproportionally big brain, compared to many other mammals. This means that birth has to take place even earlier, and that the newly born is even more helpless. Infant child rearing is central in human culture, and this is playing a role in early transmission of culture and cognitive skills while learning physical skills.

Some more examples of the co-evolutionary interrelationships will be discussed in this chapter, focussing on the key factors of interaction (tool manipulation, making, and design) and communication (language).

TOOLS AND LANGUAGE

Neurologist Frank Wilson in his book *The Hand – how its use shapes the brain, language and human culture* [1998] sees the hand in an interdependent context of biomechanical, physiological, and neurological aspects, as well as presenting the anthropological, historical, and evolutionary development of the human hand.[5] As the subtitle of Wilson's book indicates, there is a strong relationship between human cognitive development and physical development, and that technological advances are part of human cultural developments. As Wilson puts it, "No credible theory of human brain evolution can ignore, or isolate from environmental context, the co-evolution of locomotor, manipulative, communicative and social behaviours of human ancestors."[6] Wilson also points out that studies of brain mechanisms to do with language, and those to do with skilled hand use, are "segregated by discipline".[7]

Palaeoanthropologist Steven Mithen, Professor of Archaeology at the University of Reading, in his book *The Prehistory of Mind – a search for the origins of art, religion and science* [1996] presents his studies of the development of cognitive functions from early humans to modern humans. In his later book *The Singing Neanderthals – the origin of music, language, mind and body* [2005] Mithen not only presents the deep historical interrelationship between the human development of music and speech (which will be further discussed in Chapter 6, on communication), but also emphasises the reciprocal relationship between environment, tool use, and language development. In short, according to Mithen and others who have studied human evolutionary development, early communication (leading to language as we know it now) was *holistic, manipulative, multimodal, musical,* and *mimetic,* as in the acronym Mithen uses, *Hmmmmm.* This concept of *Hmmmmm* that Mithen develops throughout his book *The Singing Neanderthals* summarises this type of communication as a kind of proto-language. This is discussed in detail throughout this and the next chapters, but it is useful to have a short summary of this *Hmmmmm* concept here.

Holistic means that utterances don't have grammar, but that each concept is represented in a unique utterance. Our current use of language is compositional, also called 'atomic'; we can express many thoughts or concepts with just a limited set of elements ('atoms'), such as the phonemes of spoken language, or the letters of the alphabet in writing, or the words – these elements get modified and combined (composed) according to the rules of grammar. This is a common idea in the study of language development such as the work of Alison Wray and others that Mithen cites on proto-language.[8]

Manipulative emphasises the relationship between tool use and communication, that through manipulation of objects our language development and other cultural practices came about.

In the acronym *Hmmmmm* there are two *m*'s, for **multimodal**, the use of multiple modalities in communication, describing the strong link between the different expressive modalities, such as speech, gesture, facial expression, and posture. This is still the case, we usually express ourselves in many of these modalities simultaneously, which is further discussed in Chapter 6 *Communicating* and Chapter 11 *Expressing*. It is very likely that the multimodal forms of expression co-evolved over time.

The next *m* is for **musical**, and this is the main thesis of Mithen's book: the notion that in early communication, language and musical expression were closely related. Where in cognitive psychology and linguistics, music is often seen as an evolutionary by-product of language development, Mithen and others make an argument for a co-evolution or even that musical abilities precede language development, putting music in the forefront instead of ignoring it. Mithen admits that in his earlier book *The Prehistory of Mind* [1996] music is not considered,[9] which he says was often the case in palaeo-anthroplogical literature of that time, as well as often in cognitive science. Mithen specifically refers to Steven Pinker, who in his book *How the Mind Works* famously dismissed music as a "pure pleasure technology", using the phrase "auditory cheesecake".[10] He sees music as a nice by-product of language (exaptation, using an evolved adaptation for a different purpose). Pinker doesn't seem to be entirely unappreciative of music, but still, the attitude is provocative in two ways: firstly as an affront for anyone who regards music as one of humankind's greatest achievements (such as myself, and any professional musician or composer), and secondly in denying the important role music has had in human evolution of expression and social interaction. Mithen convincingly shows in his *The Singing Neanderthals* book that music is "more than cheesecake".[11] Henkjan Honing, Professor of Music Cognition at the University of Amsterdam, in his book *Iedereen is Muzikaal* (which is about the widespread human musical abilities, first published in 2009), refers to Pinker's cheesecake too.[12] Honing, from his music cognition perspective emphasises music as 'play' and certainly important for human communication and social interaction. Humans always have been very social animals, and music will have played an evolutionary role in establishing and maintaining social bonds. This is why singing together is still a strong form of bonding, with meditative qualities, the importance of which has been argued by biologist Rupert Sheldrake.[13] Singing and musical sound making is an activity that people can perform together, at the same time, whereas talking relies on turn taking to be effective – multiple people talking at the same time does not lead to a shared communication,* but singing together amplifies the message and the effect. One of the unfortunate side effects of the COVID-19 pandemic in 2020 was that singing

---

* As convincingly shown in the first 2020 US presidential debate, where the incumbent president regularly talked over his challenger's speaking.

together was discouraged or even prohibited, as this activity could be transmitting the virus.

And finally, the last *m* in *Hmmmmm* is for **mimetic**, because the expression would have been based on mimicking phenomena from the real world, for instance imitating an animal in posture or its sound. This will be further discussed in Chapter 6 on Information and Representation, but at this point it is important to note that this is a core concept from the field of semiotics and indeed all communication theory and interaction theory: the **signs** that refer to an object or idea are on a continuum of expression from mimetic or **iconic** representation (signs resembling the object or concept) to fully abstract, arbitrarily chosen **symbolic** representational signs.

EVOLUTION AND HUMAN DEVELOPMENT

Though no proof has been found yet, it is generally assumed that humans evolved from a common ancestor shared with the chimpanzees, around six mya. It is important to remember that this common ancestor can't be personified, other than for rhetorical purposes, as Yuval Noah Harari does in his book *Sapiens*: "Just 6 million years ago, a single female ape had two daughters. One became the ancestor of all chimpanzees; the other is our own grandmother."[14] The statement is appealing; Harari's book is full of original insights and presents a macro-vision on human development. However, as the philosopher Daniel Dennett points out, it is actually not possible to pinpoint the appearance of a new species in evolutionary processes, due to the gradual nature of these processes. He illustrates this with the hypothetical "first mammal".[15] I am emphasising this as it will be a recurring issue in this book, when discussing existing frameworks and taxonomies, and presenting new ones. "We should quell our desire to draw lines", Dennett insists.[16] Often the lines are arbitrarily drawn, and at best identify extremes on a continuum. Looking for the gradual differences on continuous scales is more sophisticated and more useful than creating (often, indeed, false) dichotomies. It seems more interesting to look for the elements between the neat divisions, which challenge some of the existing structures, and engage with these in the context of a whole spectrum.

As a case study we can look at the echidna or the platypus, which are examples of animals that have traits of different lines in the classification system and therefore challenge the human-made taxonomies. Evolutionary biologist Stephen Jay Gould describes the history of the discovery of these animals in two chapters in his 1991 book *Bully for Brontosaurus*, on the platypus and the echidna respectively.[17] The echidna and the platypus are monotremes; they have a cloaca like birds and reptiles, one opening for digestive, urinary, and reproductive purposes. Since its first discovery in Australia at the end of the 18th century, the platypus caused great confusion because of its mixed characteristics, a 'melange' usually found in different classes in the taxonomy: a bill like a bird, fur like a mammal (and milk glands), egg laying like a reptile, a tail like a beaver, webbed feet like an amphibian, no teeth, and the males have a venomous spike on their hind legs. It was first thought to be a hoax, when first the drawings and later taxidermic animals were sent back to England for detailed study, and uncertainty about its characteristics continued for a long time – it took over 80 years before its eggs were actually found (in 1884) and the theory of the animal's oviparous nature confirmed. Before that, others had suggested that it was viviparous, giving birth to live young which goes with the milk glands like any mammal, or at least ovoviviparous, meaning the eggs would dissolve within the body. This was the theory held by Richard Owen, the anatomist and later founder of the Natural History Museum in London, and opponent of Darwin. (Owen, as palaeontologist, is also known for coming up with the name dinosaur, meaning 'terrible lizard', which Bill Bryson in *A Short History of Nearly Everything* describes as "a curiously inapt name" as they are both not all terrible (only some of the famous ones) and certainly not lizards.[18])

Creationism in traditional religions assumes species to be fixed, and while the scientific approach to creating taxonomies (such as the Linnaean taxonomy of living things, but also including extinct species) allowed changes to be identified over time (evolution), there was a tendency to view this as a fixed system. Wandering around a museum for natural history (such as the Australian Museum in Sydney) one understands why a fixed taxonomy seems to make sense, through the way the 'specimens' are presented. The collection shows a snapshot of the 'natural history' (a term that in itself indicates stasis), and Owen may be forgiven in opposing Darwin and his ideas on evolution at the time. Owen believed, like many in his time, that species were fixed. The species were sampled, collected, and catalogued, eventually publicly exhibited in London's Natural History Museum. The museum, which was opened in 1881, is a magnificent building with (not coincidentally) an appearance somewhat like a cathedral. Owen actively lobbied against the placement of a statue of Darwin, which was unveiled in 1885 (three years after Darwin's death), in the museum. However, Owen's own statue was posthumously placed in the prime spot of the museum where it "commands a masterful view from the staircase of the main hall" as Bryson puts it,[19] until 2008 when Darwin's statue replaced Owen's. The inescapable symbolism of this ultimate move "to the centre stage" is shown in a BBC documentary from 2009.[20]

To return to the story of the platypus and the trouble with taxonomies, the monotremes were 'shoehorned', as Gould calls it,[21] into the class of Mammalia – where we also find another surprise from Down Under, the kangaroo, in a group of the marsupials (Metatheria), and the largest group, the placentals of course (Eutheria). As Gould somewhat sarcastically puts it, "Nature needed clean categories established by divine wisdom."[22] Or, as primatologist Frans de Waal put it, "our desire for sharp divisions is at odds with evolution's habit of making extremely smooth transitions [...] like trying to find the precise wavelength in the light spectrum where orange turns into red."[23] And to turn that metaphor back, recently it was discovered that the fur of the platypus changes colour in ultraviolet light (from brown to green).[24] This biofluorescence is common in certain plants but only recently discovered in animals, and is thought of to play a role in communication among themselves or as camouflage.

In the latter half of the 19th century it was common to interpret evolution as a linear process, probably inspired (held back, rather) by the preceding religious ideas towards greater complexity and sophistication, cumulating in *Homo sapiens*. Darwin himself, and his proponents such as Thomas Henry Huxley, never suggested this view. Darwin's sketch of the 'tree of life' in *The Origin of Species* shows many branches, and as Stephen Jay Gould advocates throughout many of his writings, the evolutionary development looks more like a bush rather than a ladder – it is not a linear process towards an inevitable goal, but a branching and meandering complex full of redundancy, still ongoing.

The linear approach, however, is shown in the treatment of the platypus and even more with another monotreme newly discovered in Australia, the echidna. Ever since they were first studied in the 19th century, the argument was made that they were 'primitive' versions of mammals, arrested in their development from bird to mammal. Even the name in many languages refers to its bird-like features: its official Latin name is *Ornithorhynchus anatinus*, 'vogelbekdier' in Dutch, and in English commonly known as duck-billed platypus. Gould shows how the platypus muzzle is in fact a mammalian adaptation and impressively sophisticated instrument for detecting prey with various very sensitive receptors. He describes the attempts to present the monotremes as primitive in detail and with almost palpable indignation in the chapters.[25] This disdain continues until today; dictionaries commonly define the monotreme as 'a primitive mammal', and even Attenborough himself uses this phrase in his documentary on Darwin mentioned above (the footage of the platypus is from 1979 but the narration is from 2009). Another recent example from the 1980s is Robert Hughes, in his otherwise very well informed and highly critical book on the colonial history of Australia, *The Fatal Shore* [1986], where he mentions that the platypus "was too primitive to regulate its own temperature". Indeed it wasn't until the 1970s that it was found that it *can* regulate its temperature as well as any mammal.[26] This should not have been a surprise, as the platypus can be found in the ice-cold water of Tasmania in winter, but as often, pre-conceived ideas and cultural attitudes can cloud the interpretations.

About the echidna, it is known that they have a large prefrontal cortex (the part of the brain usually thought of as the decision-making and cognition part) relative to body size, being 50% of the cerebral cortex and larger in fact than humans (29%).[27]

Instead of evolution being "a ladder towards progress", it "proceeds by branching", as Gould wrote.[28] Gould convincingly shows how the platypus is a highly advanced animal, superbly adapted for life in streams, and possessing a range of amazing features (its muzzle which it uses for finding prey is super sensitive, with both tactual receptors and electroception, for instance). Both platypus and echidna have an appropriate brain size and complexity.

Similarly, the evolutionary family tree leading to our own species, *Homo sapiens*, is not linear, as was often thought in the past, but has many different branches, dead-end paths, and false starts, which, as argued above, is common in evolutionary development. In *The Prehistory of the Mind – a search for the origins of art, religion and science* [1996] archaeologist and palaeoanthropologist Steven Mithen discusses this. Since this book appeared, many more of the recent branches have been discovered, such as the Denisovans and the *Homo naledi* (discussed below). An article in *Scientific American* in 2018 gives an overview, showing that archaic *Homo sapiens* was not alone, but lived simultaneously with as many as six other hominins (there is genetic evidence of this too).[29]

The first hominins that have shown up in the fossil records ("it isn't easy to become a fossil", as Bill Bryson puts it in *A Short History of Nearly Everything*, explaining the whole process and the unlikelihood of such an event taking place at all[30]) were the australopithecines or australopiths ('southern apes') such as the *Australopithecus afarensis* which lived around three to four mya. They lived mostly in trees, were able to walk upright, and used tools. The first species of the genus *Homo* appear from about two mya, such as *H. ergaster*, *H. habilis*, and *H. erectus* (and eventually *H. neanderthalensis* and *H. sapiens*). Mithen and others call them the 'early humans', and these early humans were using more advanced tools and cultural practices than the early hominins, and at least some form of language. The earliest presence of *H. sapiens* is generally assumed to have been around 200,000 years ago, though there have been finds from 300,000 years ago. Mithen calls these "early modern humans", also known as 'archaic *Homo*' or 'archaic humans'.

Homo floresiensis

A relatively recent find is that of the *Homo floresiensis* in 2004, named after the location where the remains were discovered, the isle of Flores in Indonesia. They were little people, and shrinking is a known evolutionary response to the isolation of an island ecosystem, also seen in the occurrence of for instance pygmy elephants in such situations. The Flores hominins probably descended from *H. erectus* who had spread from Africa into Asia and at least as far as the island of Java (to the west of Flores) and possibly even further east around a million years ago. In addition to being discovered so recently, what makes the *H. floresiensis* fascinating is the possibility of the relatively recent time of their extinction. At the time of discovery it was thought that this was as recent as 12,000 years ago. That would have meant that this descendant of *H. erectus*, from the first wave of hominin colonisation of the Indonesian archipelago, co-existed with *H. sapiens* (who had made it to the Australian continent by then). Indeed there are apparently many stories in local Indonesian folklore about little people, echoing their possible presence. The news was announced around the time I was working on my PhD thesis [2006] in the Netherlands, and it really made me realise how much is still unknown about our origins, and how much remains to be discovered. Recent geological studies, however, of the strata of the finds on Flores, date the chronology between 100 and 60 kya (thousand years ago).[31] There have always been disagreements and alternate views regarding the *H. floresiensis*; a good overview is given by archaeologist Josephine Flood in a recent edition of *Archaeology of the Dreamtime – the story of prehistoric Australia and its people*.[32]

Homo naledi

Another recent find was presented in 2011, of a new species of hominins based on excavations in a subterranean cave in South Africa, named *Homo naledi* (after the name of the cave, Naledi, which means 'star' in the local Sotho language).[33] This is interesting for several reasons. First of all, it again shows how incomplete our knowledge of the past is, basing our insights in human evolution on only limited fossil finds, a process that is still going on. The techniques for dating and interpreting the fossil finds are very diverse and many are very sophisticated, combining insights from biological, cultural, geological, and technological fields among others, emphasising the complex interrelationships. This inevitably leads to controversies based on the range of interpretations, some minor, some major, around earlier finds, particularly the *H. floresiensis*. Probably in order to avoid much of this confusion, this time the find and excavation of the *H. naledi* site was undertaken and presented by a large number of researchers from a range of institutions, publicly sharing their data (even 3D models of some of the remains). It is clear a strong effort has been made to strive for consensus and sharing rather than competing for prestige. The *Homo naledi* finds weren't dated when first presented, but they are thought to be habiline (such as *H. habilis*); in many ways they seem to sit somewhere between the Australopithecines and early humans. Recent research in analysing sediments and flowstones from the cave shows however an age of 414–236 kya.[34] This would indicate that this species which, based on their morphology, is thought to have evolved millions of years ago, lived until this relatively recent time. The number of bones found is unusually large, which enabled the investigators to analyse the morphology and behavioural traits. The brain size of *H. naledi* is about 500 cc, small compared to about 900 cc of the *H. erectus*. Among the many bones found is an almost complete right hand which has been analysed for functional capabilities. The thumb is extended, and overall the hand structure gives evidence that *H. naledi* would have been capable of more complex object manipulations, while still having long and curved fingers indicating that they were good climbers.

Homo denisova

The earlier discovery of the Denisovans (thought to be related to the Neanderthals) is, in stark contrast to the *H. naledi* find, based on just a few fragments, through genetic analysis. Most finding spots are in warm climates in Africa and Asia, but the temperature of this location (in the Denisova Cave in South Siberia, near the Chinese border) is around zero degrees, which means the DNA could be preserved. A team analysed the mitochondrial DNA extracted from a bone fragment (the distal phalanx of the little finger, a very modest find) and were able to infer that Denisovan DNA shows up in the genetic profile of some modern humans.

It is thought that this introgression (transfer of genetic material between species) has facilitated the adaptation of early modern humans to the demanding environments they moved into – the modern Tibetans' ability to function at high altitudes is attributed to this.[35]

Homo luzonensis

An even more recent discovery of human relative, presented in 2019 in an article in *Nature*, was of a species called *H. luzonensis*, named after the island of Luzon in the Philippines where the remains were found.[36] The find was dated at 67,000 years ago, and the hominins might have lived until as recently as 50,000 years ago. But what is even more remarkable is that the species might have been there as early as 770,000 years ago. It is not clear yet if they are a separate species of *Homo*, they might have descended from *H. sapiens* but that would put pressure on the theory of when we left Africa, or another possibility is that they are related to *H. erectus*. This species is thought of as also having been subjected to 'island dwarfing' like the *H. floresensis*. As a commenting article in the same issue of *Nature* puts it, "our picture of hominin evolution in Asia during the Pleistocene just got even messier, more complicated and a whole lot more interesting."[37]

Homo sapiens

Learning more about previous hominins, and other animals, might indeed eventually make us 'wise', as in the meaning of *sapiens* in Latin. It is clear that there is still a lot we don't know about our ancestry, and it is certainly clear that we are not the necessarily brilliant cumulative outcome of a linear development but that the tree of life is more like a bush, quite a jungle. It is good to bear in mind that our hominin ape-like ancestors didn't *en masse* jump out of the trees to go and live in the savannahs to invent all their tools overnight, but that this was a process of millions of years with gradual modifications and adaptations. The pace of biological evolution is very different to our intuitive awareness of time scales, and includes many dead-end developments, while others are more incremental developments. Now that it becomes more widely accepted that we might not be that different from the other apes, it becomes more acceptable to discuss the recent fossil finds in this "gigantic bush that is our ancestry", as Frans de Waal put it. De Waal emphasises that *H. naledi* has many ape-like features (particularly a relatively small brain size, similar to other finds such as the 1.8 million-year-old skulls in Dmanisi in Georgia), and that it seems a bit contrived to try to fit it in a linear human lineage – the approach of *hominisation* or *anthropogenesis*.[38] Similarly, another recent find from 4.4 mya of a hominin even older than the Australopithecines is the *Ardipithecus ramidus*.[39] This 'Ardi' challenges the notion of a linear development even further, as Frans de Waal points out; it has more characteristics of the bonobo, which compared to the chimpanzee is a more peaceful and gentle, matriarchal ape species.[40]

EVOLUTIONARY CATEGORIES OF TOOL USE AND COMMUNICATION

The earliest examples of dedicated stone tool use and making can be found from 3.5 mya. Object handling and manipulation techniques have been studied also in primates such as chimpanzees, to compare and better understand the abilities of the early hominins. Physical anthropologist Mary Marzke discerns four stages of development of manipulative skills in relation to tool making and use, related to an overview of various grip types based on (and extended from) the work of John Napier in the 1950s.[41]

It is appropriate to use these four stages because they emphasise interaction (both through direct interaction, as in the physical manipulation and transformation of materials, as well as the role tool use plays in co-evolution) and because these interactions have led to the way we, modern humans, manipulate and relate to physical tools and objects.

Marzke defines Stage 1 as occasional use and modification of natural materials for a purpose, and this is what other primates, as well as early human ancestors, have been doing for several million years. Chimpanzees using sticks to harvest termites, beavers building dams, platypuses building burrows, and birds and insects making nests are all examples of animals using materials as tools and as artefacts. In Marzke's Stage 2 a wider range of objects are handled, due to human morphological developments which led to increased dexterity and grip types of the Australopithecines such as *A. afarensis*. Stage 3 shows manufacture and control of monolithic stone tools (such as the hand axe), showing signs of design and planned making such as those exhibited by the early humans of the *Homo* family. Stage 4 includes assemblies such as hafted stone tools (such as axes with a wooden haft, or spears with stone points) and small objects only manipulable with a grip facilitated by shorter fingers and opposable thumb (pinching an object between the pads of the thumb and index finger). These tools were certainly designed, and used, in a planned fashion by the early modern humans such as the Neanderthals and *Homo sapiens*.

It is important to note that these stages overlap in time and location, and that they often existed (and still exist to an extent!) concurrently. There is a lot of variation depending on geographical location, due to environmental and cultural factors.

In the overview below, I tried to identify the different types of tools used, in relation to development of culture and communication. Because these types are interrelated, and include a number of interdependent developments, I have tried to give them a descriptive name, avoiding a bias in numbered stages. These names are a compromise of what may be considered a key element of each phase, but of course not the only one.

I named the tool types after the developmental phase of tool use, making, and manipulation, as this is the most relevant factor (or the most well preserved in the fossil records and surviving artefacts) and most important for understanding interaction. However, this tool development is strongly interrelated with biological evolution, cultural development, and developing communication and language abilities. The four tool types (or phases) are: *opportunistic tool use*, *dedicated tool making*, *designed tools*, and *composite tools*.

I preferred to base the tool types on Marzke's work, as her approach emphasises the interaction with the tool (and the co-evolution), rather than on other classifications which focus more on the tool properties. For instance, the five 'lithic modes' are descriptions of stone tools established by Cambridge academic Grahame Clark in the late 1960s. These are described in recent work of palaeoanthropologist and 'expert flintknapper' John Shea, who extended and restructured the work of Clark, and presents a framework of nine modes (indicated with the letters A to I) which takes into account local and temporal variations [2013]. However, these modes focus on stone tool making, which is useful, but our focus is on interaction and co-evolution, so it is more appropriate to the four broader categories of using, making, designing, and composing.

These four categories of object handling and making are then further extended with the notion of (passive) mechanical technology, which includes transmission and conversion of kinetic energy, and – later during industrialisation (discussed in Chapter 5) – active mechanical technology with the inclusion of external sources of kinetic energy (which is relevant for the way we interact with the object/system), electrical, analogue electronic, digital electronic technologies, and eventually interactive (algorithmic) systems (discussed in Chapter 6).

Opportunistic tool use

The first type of utilisation of artefacts is the phase of opportunistic tool use. Animal habits of creating artefacts, such as bird's nests, beaver dams, chimpanzee ant sticks, bower bird nests, and termite mounds are presumably older than human tool use, shown by these habits still being practised today. Many scholars in this field who study the development and abilities of early humans have indeed also studied primate and other animal behaviours.

Comparing to the developmental stages as proposed by Mary Marzke discussed above, we could identify this as Stage 1, which is the unplanned, opportunistic tool use. This has been practised by hominins since the Pliocene epoch (the geological period from over 5 mya to about 2.5 mya). Food gathering was probably also more opportunistic, finding fruit and scavenging, rather than hunting which would require more elaborate tools.

In this phase we can also identify communication between animals, and we can assume our primate ancestors too, as a form of direct use of a set of known signals but not language as we know it now. The development of language has gone through several stages and includes many aspects which will be discussed throughout this and the next chapters.

Dedicated tool making

In the next phase, which roughly coincides with Marzke's Stage 2, we see a combination of dedicated (shaping but not designing) tool making starting in the Pleistocene epoch (from about 2.5 mya to about 12 kya).

It is important to point out that there are examples of dedicated tool making in other animals, even though this is limited. The principle seems more widespread than earlier thought, as ethologist (studying animal behaviour) Frans de Waal points out for instance in his recent book *Are We Smart Enough to Know How Smart Animals Are* [2016]. It is known that animals do use tools that require planning ahead. Among the many examples in De Waal's book, is one of a female bonobo who carries a 7.5 kg rock (as well as her child) on her back, and during her about-ten-minute-long walk picks up some nuts, which she cracks using the rock when reaching her destination – a large rock surface that is used as an anvil.[42] This behaviour clearly required planning ahead. There are other examples, of octopuses in the wild who collect coconut shells (cut in half by humans and discarded) and carry them around with the intention of creating a sphere to hide in,[43] and chimpanzees who collaborate and plan the use of objects to reach food, or even an escape. In an even more recent book, *Mama's Last Hug – animal emotions and what they teach us about ourselves* [2019], De Waal gives many examples of animals with planning abilities, empathy, and sentience.[44]

In the human evolutionary development stage of dedicated tool making, the Australopithecines of about 3.5 mya most probably used fire for warmth and protection, were at least partially bipedal (walking upright part of the time but still habitually walking on all fours, and climbing in trees), presumably possessed communication abilities that surpass any animal communication but mostly based on mimesis, and had a brain size of about 400–500 cc. Cranial capacity or brain volume is often represented in cc, cubic centimetre, but sometimes the unit gram is used for weight. I prefer 'cc' as it reminds me of the way the engine size of motorcycles and cars is expressed (actually it is the volume of the cylinders, in which the combustion takes place). It is also common to illustrate cranial capacity with references to fruit and vegetables, for instance an orange-sized brain, or grapefruit, cabbage, etc., but I'll stick to the 'cc'. Our cranial capacity is about 1500 cc (comparable to the engine size of a modest car). And even more so than with engines, the cc-value is barely an indicator of performance, even though it is often thought of like that. In the brain, it is the number of neurons, and particularly the number of synaptic connections between the neurons – this is where the modern human brain really shows its complexity (as will be discussed in Chapter 7, *Being*) but also why birds with relatively tiny brains can display impressive cognitive abilities (Frans de Waal's books have many examples [2016, 2019]). The cranial capacity of the Australopithecines of 400–500 cc is about the size of a small motorcycle engine.

Particularly interesting is *Australopithecus afarensis* because some reasonably complete remains have been found. The first discovery was in 1974, in Ethiopia, of bone fossils comprising 40% of the skeleton which enabled detailed study of its anatomy. This find is popularly known as Lucy,* though the technical term of this find is A. L. 288-1.[45]

The musculoskeletal system inherited from living in the trees has proven to be a good base for dexterous artefact handling, eventually leading to the ability

---

* The name was inspired by the Beatles song the archaeological team led by Donald Johanson (with one 's') was listening to at the time, *Lucy in the Sky with Diamonds*. The psychedelic nature of the song has always inspired a common interpretation of the acronym spelt by the title, but apparently it was based on a drawing that Julian Lennon had made of his classmate Lucy and showed his father, who then wrote the song. The name then lived on as the heroine of the 2014 science fiction/action movie *Lucy*, written and directed by Luc Besson, in which Scarlet Johansson (double 's') plays the title character, who through accidental absorption of an overdose of a cognitive enhancement drug obtains superhuman powers, becomes capable of psychokinesis and space-time travel (credibility was clearly not Besson's objective but that is common in this genre), and eventually morphing into some kind of supercomputer (handing over all knowledge in the universe fitted on a large USB stick, maybe just containing the number 42). Lucy the australopithecine is present at the beginning and end of the movie, both as the fossilised specimen in the museum (Lucy's not-too-bright boyfriend describes this as "the first ever *woman*") and a CGI-rendered 'original' which seems quite faithful – the details discussed above can be seen in the closeup of the hands touching (of both Lucys, the transhuman one and the Australopithecine. This seems a homage to the Dawn of Man scene in Stanley Kubrick and Arthur C. Clarke's *2001 – A Space Odyssey* movie from 1968, turning their problematic proposal with its infinite regress (the question of who made the monolith from space) into the common time travel paradox which is just as problematic. There is a similar *2001* theme in Ridley Scott's impressive movie *Prometheus* from 2012 which explores this issue of infinite regress unintentionally as one of its many plot holes, and a further twist in his 2017 movie *Alien: Covenant*.

to throw objects overhand, and hold objects with greater precision which is instrumental for the development of tool making skills, due to the opposable thumb. A good example of such an adaptation is the development of the hand of these early hominins. The *Australopithecus afarensis* (Lucy) has thumbs which are longer than the ones in previous species. This, and further details of the anatomy of their hands, as studied by Mary Marzke, would have enabled them to touch the fingertips of the index and middle finger with their thumb and apply new grip types such as the "three-jaw chuck" – holding an object using the thumb, index, and middle finger – but not the fourth and fifth finger, like we can.[46] Although still at least some of the time walking on all fours (knuckle walking), they were certainly capable of walking upright,[47] and the thumb development of Lucy would have allowed this species to manipulate objects in new ways, possibly even being able to throw objects overhand.[48]

The early hominins, the human-like primates, used stone tools which they shaped so that they could be used for cutting and shaping materials. This is the dedicated tool making phase, by the early hominins such as the *Australopithecus afarensis* and the early humans *Homo habilis* and *Homo rudoflensis* (sometimes referred to as habilines). We could say that the Australopithecines and the habilines were the first ergonomists, fitting the task to the (early) human, by shaping one part of the artefact (the stone) for the task: a sharp edge for cutting, while another part of the artefact was shaped to fit the human hand.

Although intentionally shaped for a purpose, these tools were relatively primitive, and as Steven Mithen and others have pointed out, they seem to have been made without a 'design' in mind.[49] Design is an act of planned behaviour based on past experience and projecting future use, by the maker or by others. The shape of the raw material (say, a pebble stone) resembles the final product with modifications, and although this process of modifying is dedicated, it is not 'design'.

This tool making culture is sometimes named the Oldowan Industry, after the earliest and still important finding site of Olduwai Gorge in Tanzania (though it is a bit like saying that all steel comes from Sheffield, and all cars come from Detroit). And an industry it was: judging by the abundance of remains of these tools found, a large amount of tools were made.

Towards the end of this phase, with the appearance of *Homo ergaster* around 1.8 mya, bipedalism had been prominently developed, and brain size had increased to around 800 cc (similar to the engine size of, say, a Ducati motorcycle).

Designed tools

After a relatively long period of little brain size development, stagnancy in tool making techniques, and presumably not much change in communication skills (the ever recurring interrelationship) from 1.8 to 0.6 mya, a third phase can be identified, of designed tools. This is the time of the appearance of *Homo erectus* around one mya, with a brain size of 800–1200 cc, and later (but independently from *H. ergaster*) *Homo heidelbergensis* around 400,000 years ago, with a brain size of 1100–1450 cc (comparable to a Harley Davidson).

In the historical tool stage it is very likely that fire was used for cooking and therefore obtaining more nutrition from meat and vegetables, as will be discussed in a separate section below.

In this phase we can see tools emerge with a sophistication that indicates planned manufacture (design) and use. These are sometimes called the Acheulian tools, after a main finding site in St. Acheul in France, as described in Marzke's Stage 3. By this stage, the early humans had spread out of Africa and reached as far as the Indonesian island of Java, where remains of *H. erectus* from 1.6 mya have been found.

The Acheulian tools are much more refined than the earlier tools, which required more dexterity in both the making as well as in using. Their design reveals that the making of them was a planned activity, and that there was cultural transmission of knowledge about tool making across a large geographical area and over a long time (more than a million years). This makes this type of tool the longest lasting single artefact in history, which although a sign of stagnation in development could be seen as a factor of a successful design![50]

These artefacts also show a high degree of symmetry, and more generally a focus on aesthetics. Combined with the fact that the artefacts are often found in great abundance, and in many cases seem unused as a tool, or in a shape and size that they actually cannot be used as a hand tool, has led to speculations that they had an ornamental function. Or even that they were used to show off to others, such as potential mates. The "sexy hand axe theory", proposed by Steven Mithen and evolutionary biologist Marek Kohn develops this idea.[51] Mithen and Kohn are likening it to the biological principle that Amotz Zahavi in the 1970s called the "handicap principle", such as the peacock's abundant plumage and particularly the impressive tail feathers, to impress others. Another example of 'showing off' in the animal world is the male bowerbird's use of blue found objects (nowadays often from human making) that he uses to create an aesthetically pleasing environment to impress the female (as described in the semiotic context[52] in Chapter 6 *Communicating*). In nature these sorts of extensions are also used as an attempt to scare off predators. The handicap principle can be seen in behaviours too, such as the antelope's 'stotting' (also called 'pronking'), which are the seemingly unnecessary and sudden high jumps the animal makes when chased by a predator. It is as if by stotting the antelope expresses 'don't try to catch me, I am too agile', and indeed it has been proven to be an effective strategy, even though it seems to be wasting their energy – but it has to be true, it can't be faked. This 'costly signalling theory' only works if the actor, such as the stotting antelope, makes a real effort in the display of strength, and takes a significant risk.[53]

Mithen points out that we actually don't know if the artefacts were produced by males or by females, and although it is generally assumed it was the men who made them, there is actually no evidence for this.[54] Similarly, it has been long assumed that men were the hunters and women the gatherers, but this may be primarily based on prejudice. Recent research, for instance on the role of women hunters in the prehistoric Americas[55] and insights from traditional Australian Aboriginal culture,[56] shows a more even distribution of tasks.

To explain the abundance of objects, and their often pristine state, there is also a view that these finds are actually the *residues*. The object could have been used as a source for flakes, and that these flakes produced are the actual artefacts and not the core.[57] This would explain why the stone (core) is unused. But it is known that both techniques exist. Another theory is that the artefact wasn't a hand axe at all but more like a discus or frisbee, used to be thrown at animals. Though unlikely to be an effective strategy for hunting prey, it could have been used to deter or at least discourage predators (a function similar to our current use of barbed wire and electric fences). Anyway it is a nice example of the complexity of interrelationships that these artefacts can be read as (any combination of) tools, weapons, and deterrents, and also as ornaments, as cultural symbols, or – who knows – they might even have been used as currency. The shape of these tools is therefore also a good illustration of a common issue in contemporary design, as a (possibly) early example of the ongoing issue of finding a balance between aesthetics and functionality.

Actually there might be another interpretation possible, quite an obvious one in a way, and that is that it may very well be that these objects were used as *memory cues*. In the book *The Memory Code* [2016], Lynne Kelly gives many examples of the use of objects as memory aids in non-literate cultures. Her examples are from later stages, Neolithic times – including architectural scale objects and structures, such as menhirs, dolmen, henges (such as Stonehenge) and the Easter Island statues, and contemporary non-literate cultures. It seems very possible that at least some of the stone 'tools' were actually objects that coded knowledge for the initiated (the elders of a community). This is discussed further in Chapter 4 *Settling*.

A good example in the category of designed tools is the original Australian boomerang. This unique and still popular tool shows a profound understanding of how aerodynamics can influence the designed shape of the object resulting in particular movement (although famous for their ability to return to the thrower, the heavier ones used for hunting did not possess this feature). The linguist Bob Dixon describes this tool and its name, derived in English from the word *bumariñ* in the Dharuk language (which was traditionally spoken in the area which is now Sydney), also to illustrate the diversity of over 250 languages (not just dialects) of the original Australians before colonisation – the boomerang is known as, to give a few examples, *irrgili, wirra, wangi, tarama, wan.gim, warungan, karli,* and *kylie*.[58]

Composite tools

The final tool phase is the most recent, in the middle and late (or upper) Palaeolithic (Stone Age) period from 300,000 years ago, with species such as the Neanderthals and *Homo sapiens* with brain sizes of about 1500 cc. This tool phase corresponds roughly with Marzke's Stage 4; the tools from this phase are quite elaborate and can consist of multiple parts. These are called polyliths (rather than polypods, which are structures that are using gravity for integrity, such as dolmen,[59] or hafted tools such as an axe with a stone cleaver and wooden handle. The size of the objects is important – these small tools reveal the ability to apply a controlled pad-to-pad pinch grip (between the pads of the thumb and index finger), a relatively recent and distinctively human evolutionary adaptation; many primates have an opposable thumb, but not the ability to touch the pads of all the fingers with their thumb. By now the hominins have a fully developed opposable thumb and the ability to touch all fingers with the thumb, which enabled a number of new grip types and a dexterity not seen before. This is a specific human quality, which we all take for granted nowadays.*

Most proof of the early occurrence of this type of tool design is inferred; actual finds of traces of composite tools are only from the last tens of thousands of years. Recently, evidence of the oldest ground-edge axe was announced, from northern Australia and dated between 44,000 and 49,000 years ago.[60] It is assumed that it was an isolated, local invention, as no traces of earlier similar designs have been found in the Asian region.

In this phase of human development, hand dexterity reached a high level, which is reflected in the tools designed and made. It also shows human knowledge of other dynamic properties in nature, such as water and air flow, the latter applied in the design of the boomerang as mentioned before. Many tools in these cultures have a symbolic meaning too, sometimes weapons and other implements of defence such as shields are painted and decorated. In some cases the artefact is entirely purposed this way, losing its original meaning (as a tool or weapon). Increasingly artefacts appear that have a purely communicative, ritual, decorative, and artistic expressive purpose, and as Lynne Kelly emphasises in *The Memory Code* [2016], as memory aids. For example, the coolamon in Australian Aboriginal culture is a sort of dish used to carry objects, and the back of this dish is covered

---

* Losing this ability is detrimental, as shown in Michael Ondaatje's 1992 novel *The English Patient* where the thief Caravaggio (a character more prominent in his 1987 novel *In The Skin Of A Lion*) has been punished for his crimes by amputation of his thumbs. This is an effective means to stop Caravaggio from further stealing, and the story shows how the loss of the opposable thumb mechanism makes him unable to perform many everyday tasks or only with great difficulty (although in the 1996 film based on the novel, Caravaggio (played by Willem Defoe) shows he still has the dextrous ability to nick small things and get up to other mischief).

in patterns with symbolic meaning. Larger sizes of these dishes, known as parraja or pitchi are sometimes used as cradles to carry babies.[61]

Further developments in tool manufacture can be seen in the first mechanical constructions, such as bow and arrow, spear throwers, as well as the use of other materials such as bone and skin. These mechanical constructions use the principle of levers, conversion and transmission of kinetic energy, which leads to the category of *passive mechanical technology* discussed further in Chapter 4.

The spear thrower, as used in traditional Australian Aboriginal technology, commonly known as *woomara*[62] (*wamara* in Dharug or Dharuk language, and *atlatl* in the Americas), uses a stick held in parallel to the spear, with both rods semi-attached at the end. When the throwing arm moves forward, the top rod (spear) is released while the bottom rod is held, creating an angle between the two which acts as a lever. This facilitates the transmission of force resulting in a greater velocity of the spear, making the woomara a very effective hunting weapon. This effectiveness is possibly the reason why the bow and arrow never became popular in Australia – it was a technology that simply wasn't needed, rather than a missed opportunity as often thought.[63]

A simpler version of this principle can be seen in parks, where people who walk their dogs use an implement to throw a ball with greater velocity.

FIRE, BURNING, AND COOKING

An early example in the category of chemical technology is the use of fire. It presumably has been used by hominins for millions of years, initially mostly for staying warm and warding off predators, and later for food preparation. The latter was a significant contribution to the diet; cooking allowed much more nutrition to be obtained from food sources, as primatologist Richard Wrangham discusses in his book *Catching Fire – how cooking made us human* [2009]. It also enabled conservation of food to some degree (and increased hygiene), possibly combined with techniques such as smoking and (possibly older) techniques such as salting and fermenting.[64]

There is no direct archaeological proof for how long fire has been used (the oldest undisputed traces date back to less than half a million years ago[65]), but recent research looks at other evidence. An example is that *H. erectus* apparently slept on the ground, related to its reduced ability to climb into trees to make nests as chimpanzees do. No other primate sleeps on the ground due to the danger of predators (except for the biggest gorillas). This strongly suggests the use of fire.[66]

It is thought that the increase in nutrition due to cooking is related to the larger brain size of the *Homo erectus* and later species of hominins. And while the hominin set of teeth always showed a diversity which reflected a diverse diet, they had evolved into the smaller molar size of the *H. erectus* around 1.9 mya, and this biological change is seen as proof of the behavioural change of cooking food.[67] This puts the recent popular claim of a raw food diet as 'natural' in a different light, as Wrangham discusses in a whole chapter in his book.[68] Or as Vybarr Cregan-Reid puts it in *Primate Change*, "One of the problems with the 'paleo' diet is that no one really knows what early hominins really ate."[69]

There is also the notion of the symbolic purpose of cooked food, something we can still see in our own culture at present. Originally it might have been strongly related to a sentiment of taming nature.

The use of fire for warming (particularly in hearths, inside dwellings) and the development of clothes allowed for covering a greater variety of climates and inhabiting different areas of the world, even in ice ages. This also led to the loss of body hair, possibly with the advantage of the loss of parasites living in the fur, and therefore decreasing disease,[70] though wearing clothes would have brought a whole new range of health risks to do with lack of hygiene. The loss of hair as a result of the application of clothing also decreased the need for grooming (cleaning each other's fur, as many animals do, which has a social function[71]), and was an essential element in the development of human communication. It may have played a role in the transition of this type of communication from physical interaction to vocal exchange, which may have already been developed due to the increase in typical group size (as early as the *H. habilis*) which would have made physical grooming impractical and therefore extended (and eventually replaced) by vocal grooming. This theory was proposed in the early 1990s by palaeoanthropologist Leslie Aiello and evolutionary psychologist Robin Dunbar, who looked at the interrelation between group size, brain size, and language evolution, to suggest that vocal grooming might have led to the emergence of language.[72] This might be too strongly put, because words would have such a different effect than physical manipulation. Steven Mithen discusses this in *The Singing Neanderthals*, but sees a potential link with the emergence of singing rather than speaking as a vocal substitute of physical grooming.[73] This type of exchange, called *phatic* communication (still widely used, in chit-chat and small talk), and the role in language development is discussed further in Chapter 6 on information and communication.

In addition to the role of fire for warming, and subsequent evolutionary changes in body hair, and cooking, fire was also used for tool making. Heat was applied in the process of preparing wood (straightening of sticks to use more effectively as spears*), as well as for stone manipulation: by heating a particular rock the crystalline structure could be changed changing its flaking qualities.[74]

---

* This skill is demonstrated in the film *Ten Canoes* by Rolf de Heer from 2006, in which contemporary Australian Aboriginal actors present their traditional story that involves the creation of straight spears, spear throwers (woomaras), and (indeed) a number of bark canoes.

Intentional smoke signalling was also used for communication in these and other ancient cultures. For instance, in traditional Australian Aboriginal culture smoke was used to indicate presence, and in some cases inherently requesting permission to cross the land of a neighbouring group.[75] This is a good example of the distinction made at the beginning of this chapter, between an energy source that can be used for *power* (the heat of the fire for tool making and other purposes), and *communication* (through smoke signals).

To give an example of storing energy, by producing charcoal the potential energy stored in the wood is made more easily accessible for later use.

In traditional Australian Aboriginal culture the use of fire is common – it is used in order to change and influence the landscape as a kind of maintenance, effectively using fire as a farming technique. Smoking ceremonies are still common in this culture, as a cleansing ritual. This will be discussed in more detail in Chapter 4, about the development of settlements and agriculture.

BIPEDALISM AND ITS EVOLUTIONARY CONSEQUENCES

A number of changes occurred to the primates or early hominins due to their gradual transition from living in trees and woods, to open grasslands and savannahs in Africa.

Anatomically this change in living conditions freed up muscles and skeletal structures to accommodate the development of a more advanced vocal system (instrumental for the development of vocal communication such as speech), larger size of social groups (common in species that live away from the permanent cover and protection of trees) which increased the need for vocal communication, and different grips and abilities for object handling and manipulation. There are many speculations about why this change from tree living to open land living occurred, with an inevitably high degree of chicken-or-the-egg circular reasoning, but it is most likely due to changes in the environment. We might never be able to trace this back with full certainty, but it is known that the climate changed in the habitat of the hominids around that time.

Two stages of evolutionary transition

A recent insight that Mithen discusses in *The Singing Neanderthals* is that the transition to bipedalism took place in two distinct stages.[76] Each stage has its own cause or *selective pressure* – a term that is used in evolutionary theory: a change (*selection*) occurs in response to a certain influence in the environment (*pressure*). The first stage influenced the Australopithecines, possibly as a strategy for picking fruit, and led to a partial bipedalism by stretching upwards. Indeed, the early hominins were still common knuckle walkers and good climbers, but their physiology and skeleton started to adapt to a more upright posture. This corresponds with the *dedicated tool making phase* as identified above. The second and decisive stage that Mithen describes is generally assumed to be based on evolutionary responses appropriate for living in open grasslands. As mentioned before, this may have been enforced through the climatic changes to the environment, walking upright as a strategy to cool the brain as well as a better support for visual perception. By walking upright less of the body is exposed to the direct sunlight, and the air is cooler further away from the ground (kangaroos and other mammals use this strategy too). This happened during the *designed tools phase*, with the early humans *Homo ergaster* and later with *Homo erectus* (indeed).

The adaptations to the body included a better sense of balance, stability, and mechanical energy distribution during the movement cycle using strong tendons, facilitating running and jumping. All these movements however would have required much more control from the brain, and this may explain the salient increase in brain size around this time. Meanwhile, because the arms and hands were freed from their task of supporting locomotion and climbing, they could be employed in other ways. The increase in sophistication in tool making in that period is thought to reflect this increased dexterity. Bigger brains may have led to an increase in language capabilities, supported by manual activity, and the selective pressure of social groups requiring communication for complex group activities. As Steven Mithen puts it, "intelligence may have been no more than a spin-off from walking on two legs,"[77] with the development of language as a further consequence. This development was supported by the increased abilities of the vocal system owing to anatomical changes, resulting from the behavioural development of bipedalism. When walking upright the spinal cord can connect to the brain from below rather than from behind, as is the case with animals that walk on all fours. This enabled the larynx to lower in the throat, and this lengthening of the vocal tract led to a greater variety of sounds that can be produced. There is a trade-off here though, the risk of choking on food has also increased, but we seem to have adapted to that very well and this occurrence is extremely rare.[78] Furthermore, when walking upright there is less need for the larynx to act as a strong valve for chest stabilisation. This valve function of the larynx is used when holding the breath to create a stable basis for the upper limps to undertake activities that require strength such as climbing and knuckle walking.* A weaker

---

* It is therefore hard to believe that Cesar, the lead chimp, and some of the other apes in the *Planet of the Apes* remake (the trilogy *Rise* in 2011, *Dawn* in 2014, and *War* in 2017), would be able to talk like they did (it is more convincing in the 2001 remake by Tim Burton in which the apes had about 3000 years to evolve this and other cultural traits, and as such stays closer to the original story). Some apes are portrayed as using sign language, which is more credible, particularly because there have been several scientific projects undertaken in which apes were taught sign language, to investigate their language abilities, further discussed in Chapter 6 *Communicating* and Chapter 11 *Expressing*.

larynx has more vocal capabilities (as can be seen in reed instruments such as the saxophone or the hobo – the reed needs to be thin enough to vibrate under the air pressure and lip pressure applied by the player for optimal control).

The change in diet, related to the use of fire to prepare food (as discussed above) also played a role in essential changes in anatomy supporting vocal capabilities.

All these factors together can explain the emergence of the wide range of vocal capabilities of *Homo*, without the need for primary selective pressures to do the evolutionary work. "These developments begin to look like no more than an accident of evolution" is the sobering remark made by Steven Mithen in continuation of his statement quoted above.[79] This is echoing the famous phrase of the late palaeontologist Stephen Jay Gould about the emergence of human intelligence as "a glorious accident", and in line with Darwin's original ideas. In the late 19th century though, with widespread colonialism by the Western world, 'social Darwinism' became a reasoning to justify the moral superiority of the human race and particularly the 'civilised' Western world. There was no such justification in Darwin's work, these are later interpretations, nor is there any justification for assuming our species to be superior. This issue will be discussed towards the end of Chapter 4, but at this point it is good to be reminded that it is not so much about our species just being superior (as a species we have shown to possess a profound ability to wreck our habitat), but if anything our increased potential comes with the responsibility to become a species that looks after the planet, becoming true custodians.

Influence of bipedalism on rhythm and musicality

Another important element of bipedalism, as Steven Mithen points out, and this is often overlooked, is that our sense of musical rhythm is strongly related to this trait of complex coordination of walking, running, skipping, jumping, etc., in a whole chapter of *The Singing Neanderthals*,[80] 'Getting into Rhythm – the evolution of bipedalism and dance'.* Mithen argues that this could have played a major role in our development of both music and language. It is also thought to be related to the way we have always used manual tools, often through repetitive rhythmic movements as Steven Pinker has pointed out.[81] This can also explain why so few animals have the ability to anticipate rhythm, even though there are many animals with

---

* As with all the chapters in *The Singing Neanderhals*, Mithen refers to a piece of music to inspire the reflection, in this case Dave Brubeck's *Unsquare Dance* from 1961, which is deliberately cheeky as this piece is in 7/4 time signature, and while being obviously groovy is actually really hard to dance to ("tricky huh" laughs drummer Joe Morello at the end of the recording – Morello famously played a drum solo in 5/4 in an earlier unsquare piece by the Dave Brubeck Quartet, *Take Five*, composed by saxophone player Paul Desmond. *Take Five* got recycled (in 4/4!) by St. Germain's track *Rose Rouge* on the 2000 album *Tourist*).

vocal capabilities (singing birds, elephants, dolphins, bats), as studied by among others Henkjan Honing, who specialises in research on the human sense of rhythm (maatgevoel in Dutch): the ability to synchronise to a beat, as summarised in his book published in 2018.[82] We can predict and antici-pate a beat, even a salient missing beat like in a syncopated rhythm,* which Honing calls beat induction (rather than beat perception, to emphasise that we can 'perceive' a missing beat – but this is in actual fact a rather good example of the active nature of perceptual systems, which is discussed fur-ther in Chapter 9 *Experiencing*; we actually do perceive things that we don't sense, as the Gestalt psychologists have shown in many examples since the early 20th century, and worked out in J. J. Gibson's ecological approach to perception in the 1960s and 1970s[83]). The ability to pick up a beat is a fundamental element of musicality, and this has been shown to be innate – newborn babies have such an ability.[84] Honing gives an overview of the recent research involving animal abilities for beat induction in his recent book. He shows that recent YouTube hits of *Ronan the headbanging sea lion* and *Snowball the dancing cockatoo* actually have been scientifically proven to be real, and describes other recent research into the beat detect-ing ability of, indeed, monkeys. As in other examples of animal behaviour, I think these cases can illustrate evolutionary developments of species and are therefore very relevant (in addition to the importance of animal eman-cipation in general). It is also important to bear in mind that animals are of course often very able to synchronise to each other's movements and vocalisations: why would they be able to groove to human music rather than their own? Similarly, for a long time it was thought that animals were not as good as humans in facial recognition tasks – until the tests were car-ried out with faces of their own kind and they turned out to be just as good as humans.[85]

The relationship between this ability and bipedalism needs to be further investigated, but it also indicates the importance of the dancing and musical abilities that we presumably developed by using our hands (clapping) and objects from the most primitive musical instruments, another example of co-evolution.

As mentioned above, Henkjan Honing earlier (in 2009) wrote a book about the widespread ability of humans to appreciate and participate in music, "everyone has musical abilities" – even those who insist that they are 'a-musical'. Mithen's thesis that music and language have co-evolved, and might have a single origin, is nicely complemented by Honing's book which explores the relationship between music and language, and humans' musical abilities as universal but not uniform – it is strongly dependent on culture and history. Further insight is provided by neurologist Oliver Sacks

---

* *Get Up For The Downstroke*, as George Clinton and his colleagues including Bootsy Collins and Bernie Worrell have been urging us since the early 1970s (on the opening song on the second album of Parliament, *Up For The Downstroke*, in 1974).

in *Musicophilia – tales of music and the brain* [2007], referring to Mithen's work, and presenting a number of fascinating cases of people who have lost their language abilities (due to brain damage, aphasia) but not their musical abilities, strongly indicating that we have evolved different brain areas to support these skills. This research can tell us a lot about how the brain has evolved, and how it currently works, which is further discussed in Chapter 6 *Communicating* and Chapter 10 *Thinking*.

From walking to running

As mentioned before, learning to walk is one thing, but the skill of running is a further development. This can be seen with any infant, who first learns to stand up, then walk, then run, then skip, jump, dance, kick a ball, etc. (a good example of ontogeny (individual development) reflecting phylogeny (evolutionary development of the species)). A number of features related to the ability to run long distances were already present in *Homo ergaster*, including extended breathing techniques (ability to breath regularly through the mouth under strenuous activity, rather than panting) and thermoregulation capabilities (presumably the loss of hair, development of sweat gland activity, and further adaptations for cooling the body and particularly the brain). This could have been an advantage in hunting, because although those early humans were not as fast runners as their prey, they would hunt using the technique of endurance running, eventually leading to a catch.[86]

Sometimes we go back to walking on all fours, using walking sticks, which I occasionally use for hiking (and popularised as 'Nordic Walking') to take some of the load off my knees particularly in hills or mountains when carrying camping gear.

Our whole bodies have evolved to walk and run, which is good to bear in mind after all the focus on manual dexterity and our opposable thumb (as one could expect from a book called *The Hand*) and vocal capabilities (as expected from a book called *The Singing Neanderthals*). In *Primate Change*, Vybarr Cregan-Reid emphasises that our feet also had a long evolutionary trajectory to adapt to walking, sensing the surface and supporting balance. Primate feet are mainly for gripping, not so much dexterity.[87]

DISPERSAL ACROSS THE CONTINENTS

Homo sapiens developed from around 200,000 years ago in Africa (but possibly earlier, there is a find 300,000 years old), and like some earlier humans they started to move into the Eurasian continent via the Levant (now Middle East) around 100,000 years ago, and are thought to have arrived in Europe around 40,000 years ago. The migration into the American continents was much later, about 10,000–20,000 years ago, via the (then) land bridge from Asia to Alaska.

They possibly also moved from the African continent via a land bridge (sea levels were lower then) through the Arabic peninsula into South Asia (a recent find of footprints in the Nefud Desert in Saudi Arabia is dated at over 110 kya[88]). Following the coastline east, they eventually reached the Indonesian archipelago (like *H. erectus* earlier), and from there into the Australian continent. They arrived in Australia as long ago as 50,000–60,000 years, and recent finds and insights suggest even 70,000 years ago – further indication of the time when they left Africa. The Australian continent was at that time linked with what is now the island of New Guinea (together known as the Sahul continent – this connection existed until 18,000 years ago), but separated from the Indonesian islands (or peninsula rather, even Sumatra and Java were linked, for instance). Even at the sea levels at the time (60 m lower than present), it still required a sea crossing from the island of Timor (the most likely route taken) into the north of Australia. The distance of 90–100 km was such that the continent could not be seen, but its presence was probably known to the people due to the smoke of bush fires, presence of migrating birds, floating flotsam, etc. This is discussed by Josephine Flood in *The Archaeology of Dreamtime* and by archaeologist Scott Cane in the book *First Footprints*.[89] Cane presents these historical insights and combines these with his own experiences sailing and travelling in the Indonesian archipelago, including first hand encounters with the current hand-made craft that shows the (impressive) ability to make this type of sea crossing. He also takes into account the favourability of (seasonal) trade winds and tides. The handful of shorter sea crossings (up to 20 km at the time) to get to Timor would have enabled the people to develop their sea faring skills to eventually reach Australia.

DEVELOPMENT OF SYMBOLIC REPRESENTATION AND LANGUAGE

As mentioned above, in the phase of composite tool making we also find the first evidence of symbolic communication through artistic expression, the use of representation through painting and objects (mostly limited to *H. sapiens* though), and cultural practices such as burying the deceased (which the Neanderthals also did, and possibly other early humans too).

It is generally assumed that *Homo sapiens* were the first to use symbols, representing objects and ideas in abstract form, though it is hard to say as the absence of evidence is no evidence of absence (this phrase is often used in history); the use of symbols in spoken language leaves no trace. An explanation for the absence of any use of symbols by earlier hominins, or as some think, even the Neanderthals who lived concurrently, is that *Homo sapiens* might have had a genetic change, which led to an advantage in evolutionary sense through natural selection. As mentioned at the beginning of Chapter 2, there is some evidence for this theory.[90] It might have happened as early as 200,000 years ago, but we only have actual proof (however circumstantial) of symbolic expression of 60,000–70,000 years ago. This relates to the idea that our language ability is a unique property of the human brain, leading to the notion of a 'universal grammar' as has been developed by linguist Noam Chomsky since the 1960s and which can also be found in psychologist Steven Pinker's notion of the 'language instinct', the title of his 1994 book. While this nativist notion seems to make sense to some, recent insights support the realisation that language might have mainly emerged from a large range of biological and behavioural changes, some unrelated, as discussed in this chapter. Anthropologist and linguist Daniel Everett, who as mentioned in Chapter 2 studied the language of an Indigenous people in the Amazon, even goes as far as stating that language is a 'tool', just like any other tool we developed in response to certain needs and environmental pressures.[91] And the making of artefacts, as well as the increased group size of communities, would have required a lot of communication. Living in larger groups would have influenced the development of communication, due to the complex social demands and particularly the need for advanced communication around complex tasks such as group hunting and joined tool making activities. The latter, called heterotechnic cooperation, is still demonstrated in surviving cultures working with stone and wood tools such as in New Guinea and Australia[92] and is of course currently a common way of working in craft, design, and industrial manufacturing situations with its specialised division of labour (discussed in Chapter 5 *Industrialising*).

Steven Mithen's explanation in *The Prehistory of the Mind* [1996] and repeated in his later book [2005] is that previously (with the early humans) the mind was fragmented into different mentalities or knowledge domains, such as nature, technology, communication, etc. based on theories from the 1980s and 1990s about the modularity of the mind. When *H. sapiens* started to integrate these separate domains, achieving what Mithen calls "cognitive fluidity", symbolic thinking, abstract reasoning, and language would be able to emerge.

This is further discussed in the chapter on human thinking (Chapter 10), because recent insights into the relationship between rational and intuitive thinking, or cognitive versus unconscious thought, may be a more suitable way to look at this. Again, it seems entirely plausible to me that at this point in human development language and complex thinking can have naturally *emerged* from the complexity of the mind, in relationship with others (social), tools developed (technology), and the environment (nature).

Evidence of *H. sapiens'* ability to refer to objects by representation, either by mimesis (drawings of animals on cave walls – parietal art – or in the sand) or abstraction, has been found only from relatively recent sources of the last 50,000 years, such as art objects and musical instruments. Evidence from earlier times is more circumstantial, such as the finding of large amounts of pieces of ochre, presumably used for painting human bodies or objects, or stones with patterned incisions from 70,000 years ago. The search for physical evidence is still going on in Africa, but the later sources, such as the famous Grottes the Lascaux in France, the Cueva de Altamira in Spain, and other finding places in Europe are convincingly showing the development of the use of symbols and artistic expression around 20,000 years ago in France and 36,000 years ago in Spain.

When first discovered in the late 19th century, there was disbelief that Stone Age people would have been able to create such vivid, realistic, and expressive imagery.[93] This neglect of our prehistory is still continuing, though in the last decades there is new interest and appreciation (and inevitably romanticisation) of earlier cultures. Knowing where we have come from, and how our skills and knowledge have developed, is a firm basis for understanding the present, and helps anticipating future developments.

The caves were generally not used as dwellings, they were mostly used for ceremonial purposes. This may have included musical expression, as has been speculated based on the remarkable acoustic properties of many of the locations.[94] Although the imagery seems primarily mimetic, they show many stylised elements and even abstract symbols. The meaning of these abstract symbols is not known, they were often overlooked as the main focus was on the mimetic elements. These symbols are the topic of recent research by Canadian palaeoanthropologist Genevieve von Petzinger, who has visited many caves, assembled images of all abstract symbols, and organised these in a database for further study as presented in her recent book.[95]

Much older artefacts than what has been found in Europe and Africa are increasingly emerging in Australia, where the start of human presence is now dated at up to 65,000 years ago or more, and symbolic artefacts of over 30,000 years old have been identified. Since the time of first firm contact with Western culture with the arrival in 1788 of the British 'first fleet' of convicts and soldiers, this long-surviving culture which was developed over millennia has been severely damaged but there is still a lot left. After the initial misunderstandings and misinterpretation, and appalling treatment of the Indigenous population by the colonists, eventually attempts for appropriate appreciation, respect, and support for Aboriginal culture was established, particularly in the last decades.

Traditional Australian Aboriginal culture is, although confined by the tools and techniques accumulated as described above, an incredibly rich culture and was particularly well suited for the continent. Aboriginal technology and cultural practices have been studied to reveal their historical development and continuously evolving aspects, particularly their profound ability to communicate across the whole continent (discussed further at the end of Chapter 4 *Settling*).

In addition to the earlier lack of appreciation and understanding, the fact that the Australian continent is vast, spanning distances of thousands of kilometres, and only thinly populated compared to Europe, means that much of the richness still needs to be explored. There is an already impressive range of finds in archaeology of human remains and traces, such as ancient rock art paintings and engravings. Recent research increasingly pushes the evidence of human presence on this continent further back in time, and has uncovered finds of artefacts that prove that symbolic representation in the Australian Aboriginal culture is the oldest in the world.

EVOLUTIONARY RELATIONSHIPS
BETWEEN EARLY HUMANS

It is generally assumed, supported by genetic evidence, that the *Homo neanderthalensis* and the *Homo sapiens* shared a common ancestor sometime between 300,000 and 500,000 years ago somewhere on the African continent. The Neanderthals evolved in the European continent and Central Asia around 250,000 years ago while the *H. sapiens* evolved in Africa and only arrived in Europe about 40,000 years ago. Neanderthals had large brains, slightly bigger than ours in fact, but corrected for body size their so-called encephalisation quotient (EQ, the ratio between body size and brain size) is actually slightly lower. They possessed fine motor skills, and their ability to survive in Europe during the Ice Age is certainly impressive and would only have been possible if they possessed some elaborate form of communication in addition to their technological and making skills. Their communication would have been 'Hmmmmm' as Steven Mithen calls it (described above), as reflected in the title of his book *The Singing Neanderthals*, but presumably only limited symbolic language (not using many abstract signs). This theory is supported by a striking absence of almost any form of symbolic artefact, but all ingredients as summarised in the Hmmmmm acronym; holistic, manipulative, musical, multimodal, and mimetic seem to have been present.

Recent research however has shown that some of the cave paintings, particularly the symbolic (rather than the mimetic) elements, can be attributed to the Neanderthals. This is based on an improved version of an existing dating method (analysing uranium-thorium balance in calcium layers) which shows that the age of several (elements of) cave paintings in three different locations in Spain is at least 64,800 years, implying Neanderthal authorship.[96] It is interesting how new finds and insights continuously evolve the knowledge about our prehistory. This example shows how biases and (working) hypotheses are influencing interpretations. Previously it was thought that only *H. sapiens* would have had the ability for this type of symbolic (abstract) thinking, and it was concluded that they must have been the ones producing these art works. The new dating is interpreted as proof of Neanderthal ability of this kind, although in the future someone might see it as proof that *H. sapiens* arrived in Europe much earlier.

The Neanderthals probably became extinct around 40,000 years ago, according to the latest and more accurate dating of remains presented in 2014.[97] Although there is no direct evidence that this was caused by the arrival of *H. sapiens*, there is an overlap of at least several thousand years so it seems likely that there is a connection here.

Likewise, there have been debates about whether any interbreeding between *H. sapiens* and *H. neanderthalensis* took place, and only recently genetic analysis has proven that this was the case – actually as well as with some of the other early human hominins existing simultaneously (as discussed above).[98] In some cases these genetic interrelations manifest themselves, such as in the example of the *H. denisova* genes that have benefited some modern humans' ability to function at higher altitudes, and the recently discovered relationship between 'Neander-DNA' and vulnerability to COVID-19. It turns out that it negatively influences the chances of suffering from more severe effects of the disease, which has been shown in Europe and South Asia where there is some presence of *H. neanderthalensis* genes in *H. sapiens*.[99]

Our evolution looks definitely much more like a bush than a hierarchical 'tree of life', an entangled shrubbery* with many interconnections between branches rather than a linear ladder of development (Darwin actually contemplated using coral as a metaphor). Several earlier hominin species are related to the later *H. sapiens*, stretching over several hundreds of thousands of years, much further back than the traditional figure of 200,000 years, and with many more interrelationships ("an African cloud", but also on other continents).[100]

There is no reason to assume we are the unavoidable and profound accumulation of everything that happened before, and therefore no reason to claim any superiority over other species or nature in general. It is the result of an imperialistic world view, common in Darwin's time, dominant until at least later in the 20th century. The phrase 'survival of the fittest' (coined by Darwin's admirer the biologist Herbert Spencer in the years after the publication of *On the Origin of Species*[101]) emphasises struggle and battle (as suggested also by Darwin in the subtitle of the book *preservation of favoured races in the struggle for life*). To an extent this is apt, survival in nature is not trivial, there is competition for all kinds of resources, material and social. However, recent findings in biology and ecology show a stronger emphasis on collaboration rather than competition. In the recent book *Entangled Life: How Fungi Make Our Worlds, Change Our Minds, & Shape Our Futures*, biologist Merlin Sheldrake, as mentioned above in the Preface, explains the many symbiotic relationships in nature. Whereas previously many of these relationships were seen as parasitic, it is now very clear that many species actively collaborate in symbiotic relationships. Sheldrake sees it as a continuum, a spectrum from parasitic to symbiotic.

---

* *"Ni!"*

One of the earliest examples in nature are lichens, which are a symbiosis between fungal and algal partners (and therefore definitely a challenge to taxonomies), first understood as such in 1877 by the German botanist Albert Frank.[102] There are many examples of symbiosis, collaboration and interdependencies in ecosystems and even in individual organisms, if one cares to look for this – which, as mentioned, was for many years actively opposed due to the prevailing world views. The American biologist Lynn Margulis has been proposing the notion of *endosymbiosis*, multicellular life evolved from single cell organisms collaborating together.[103] This is now commonly accepted, but in the late 1960s, when Margulis first proposed it there was a lot of resistance to the idea. The symbiosis also takes place inside an organism, and our microbiome (the microbial ecosystem in our gut) is our own symbiotic partner system which influences our behaviour in many, often subtle, yet sometimes radical ways.[104] On a larger scale, there is the idea of approaching the whole planet Earth as an organism or an ecosystem, the Gaia theory, as developed by James Lovelock in the 1970s[105], further discussed in the section *Humans versus Nature* at the end of Chapter 5.

The fascinating lifeform of siphonophores are another example of symbiosis. Siphonophores consist of several seperate, specialised species, that form a kind of colony, a conglomorate organism. (The 'Portugese man o' war' is an example, it floats in the sea and can stun or kill prey with venom from its tentacles (like many jellyfish) – they are common on Australian east coast beaches and known as 'bluebottles' and have a painful sting – on one occasion I had one wrapped around my leg which was pretty nasty.)

The interdependancy between trees and fungi, as mentioned above in the Preface, is a key example in Sheldrake's book. Not only do trees and fungi exchange nutrients (which was actually already suggested by Frank in 1885), the mycorrhizal networks in the soil allow for communication between trees by chemical means (first proved in 1997 by forest ecologist Suzanne Simard, using two radioactive isotopes C13 and C14 to also reveal the bi-directional nature of the connections – in the wild[106]). This led to the term 'wood-wide web', although the most fascinating aspect is the mycorrhizal 'neural' network, more than the wood*.

It is now abundantly clear, particularly when viewed through this cooperative lens, that (endo)symbiosis, collaboration, and interdependency is the

---

* As visualised in the 3D movie *Avatar* of James Camaron in 2009, the phosphorous lighting up of the hyphae in the soil on the planet Pandora (as mentioned by Merlin Sheldrake [2020, p171]). The documentary *Fantastic Fungi* by Louie Schwartzberg from 2020 shows impressive cinematography of time-lapse, sped up footage of emerging fungi, the fruiting bodies we encounter above ground (and which are in some cases delicious to eat), which makes them look very alive. Similarly, computer generated imagery visualises the underground mycorrhizal networks simulated with pulsating lights – this is sped up too which gives a stronger impression of the 'neural network' of fungi, though this depiction is effective it is perhaps a bit inaccurate.

norm in ecosystems, rather than parasitic, competing, fighting and conflicting. It is the 'survival of the fittest *collaboration*'.

NOTES

1. Bill Bryson discusses the age of planet Earth, including the historical development of the current estimate, in *A Short History of Nearly Everything* [2003, p202]. See also *Sapiens* by historian Yuval Noah Harari [2014], which describes the emergence of life on our planet. Merlin Sheldrake in his book *Entangled Life – how fungi make our worlds, change our minds, and shape our futures* [2020] discusses what role fungi played as a catalyst in enabling plant life (and still do), and how eventually animal life also emerged from the sea.
2. [Bryson 2003, pp394–406], [Gribbin, 2018].
3. As archaeologist Josephine Flood, author of *Archaeology of the Dreamtime – the story of prehistoric Australia and its people* defines it [2010, p16].
4. *Archaeology, a very short introduction* by Paul Bahn [2012, p2].
5. Frank Wilson's book *The Hand – how its use shapes the brain, language and human culture* [1998] was recommended to me on several occasions by Bill Verplank. Wilson discussed this evolutionary and biological interrelationship in his keynote talk at the *Tangible and Embedded Interaction Conference* at Stanford University in January 2015, invited by Verplank who was the conference chair.
6. Frank Wilson, *The Hand – how its use shapes the brain, language and human culture* [1998, p321].
7. [Wilson 1998, p325n21].
8. [Wray 1998], [Zachar 2011].
9. As Mithen writes "I am guilty of this myself, having failed to consider music in my 1996 book" in *The Singing Neanderthals* [2005, p4], and "I am embarrassed by my own previous neglect of music" [2005, p5]. In the preface of the book [2005, pvii], Mithen is also upfront about having no musical skills himself, but is clearly appreciative of music.
10. [Pinker 1997, p534].
11. The title of Chapter 2 [Mithen 2005].
12. Cheesecake is 'Kwarktaart' in Dutch [Honing 2012, p21]. Henkjan Honing's book *Iedereen is Muzikaal – wat we weten over het luisteren naar muziek* (which means something like 'everyone has musical abilities – what we know about listening to music') is available in an English translation *Musical Cognition – a science of listening* published by Routledge in 2011.
13. Rupert Sheldrake, *Science and Spiritual Practices – reconnecting to direct experience* [2017].
14. [Harari 2014, p5]. A newspaper review of the book added to this "Try to stop reading after such an opening....". This review appeared in the *Sydney Morning Herald* of 22 November 2014, written by Prof. Glyn Davis, then vice-chancellor of the University of Melbourne.
15. Daniel Dennett, *Intuition Pumps and other tools for thinking*, Chapter 43, 'Beware of the Prime Mammal' [2013].

16. [Dennett 2013, p241].
17. Chapters 18 'To Be a Platypus' and 19 'Bligh's Bounty' [Gould 1991, pp269–293], originally published as articles in *Natural History* magazine, in August and in September 1985.
18. [Bryson 2003, p120].
19. [2003, p124].
20. *Charles Darwin and the Tree of Life* by David Attenborough.
21. [Gould 1991, p274].
22. [1991, p275].
23. [De Waal 2019, p45]. Ethologist Frans de Waal is Professor at the Emory University and director of the Yerkes National Primate Center near Atlanta.
24. [Anich et al. 2020].
25. [Gould 1991, pp269–293].
26. [Gould 1991, p277].
27. Tyson Yunkaporta opens his book *Sand Talk – how indigenous thinking can save the world* with an insightful reflection on the echidna, and its relatively large cerebral cortex. This fact can be found in scientific literature, but as Tyson pointed out during one of his recent lectures for UTS (on Zoom), it is rare, he expected many more scientific sources on this. Although his book often backs up claims very well, there are some omissions which Tyson eagerly admits – and this is one of them. When I offered the reference that I did find (indeed, not easy to find a good reference, though it is mentioned on the Wikipedia entry), the book *Echidna: Extraordinary Egg-laying Mammal* by Michael Augee, Brett Gooden, and Anne Musser from 2006 (it's on page 48), Tyson immediately recognised the names and realised this as his source too (an article of these echidna specialists). Yunkaporta's lecture was for the New Knowledge Making Lab course of the BCII double degree programme of the transdisciplinary faculty at UTS, on 22 February 2021.
    In the book, Yunkaporta states "Sometimes I wonder if echidnas ever suffer from the same delusion that many humans have, that their species is the intelligent centre of the universe." He furthermore emphasises the importance of echidnas as totemic entities in traditional Australian Aboriginal culture [2019, p1].
28. [Gould 1991, p276].
29. The title of the article is *Last Hominin Standing* [Wong 2018].
30. [Bryson 2003, p390].
31. [Sutikna et al. 2016].
32. [Flood 2010, p4].
33. [Berger et al. 2015].
34. [Dirks et al. 2017].
35. [Huerta-Sánchez et al. 2014].
36. [Détroit et al. 2019].
37. [Tocheri, 2019].
38. Frans de Waal uses these terms, and he also couldn't resist pointing out that *naledi* is an anagram of *denial* [De Waal 2019, p45] – a coincidence of course, as *naledi* refers to the name of the cave. This section of the book was published in a different form earlier as an opinion piece 'Who Apes Whom?' in *The New York Times* on 15 September 2015 (p23).

39. There is an article in *Science* magazine that covers the details of the finds [Gibbons 2009], and this hominin is discussed in *The Singing Neanderthals* [Mithen 2005, p122].
40. [De Waal 2019, p195].
41. Mary Marzke presented the stages in a journal article [1997]. Frank Wilson in *The Hand* covers the work of Marzke [Wilson 1998, pp24–32]. The grip types as described by Napier in the 1950s are also discussed in *The Hand* [Wilson 1998, pp118–120].
42. [De Waal 2016, pp213–215].
43. [De Waal 2016, p94]. See also *Other Minds – the octopus and the evolution of intelligent life* by Peter Godfrey-Smith [2016, p64]. (The title of the US edition is *Other Minds – the octopus, the sea, and the deep origins of consciousness*).
44. The title *Mama's Last Hug – animal emotions and what they teach us about ourselves* [2019] refers to Mama, the matriarch of the chimpanzee colony at Burger's Zoo (in Arnhem, the Netherlands) where Frans de Waal did his seminal research in the late 1970s on the complex social and political structures in the colony. As De Waal writes at the opening of Chapter 1 of his book [2019, p13], "One month before Mama turned fifty-nine and two months before Jan van Hooff's eightieth birthday, these two elderly hominids had an emotional reunion. Jan [...] is the biology professor who supervised my dissertation long ago. The two of them had known each other for over forty years." There is a YouTube video of the "last hug", showing this emotional event between species, a beautiful moment and important to reflect on.
45. [Mithen 1996, p16], [Mithen 2005, p123], [Wilson 1998, p15].
46. [Wilson 1998, pp24–28].
47. [Mithen 2005, p142].
48. [Wilson 1998, p27].
49. [Mithen 2005, p125].
50. [2005, p164].
51. [2005, pp188–191], [Kohn and Mithen 1999].
52. Dario Martinelli in the book *A Critical Companion to Zoosemiotics* [2010, pp116–119].
53. [Dennett 2013, p238].
54. [Mithen 2005, p190].
55. [Haas et al. 2020].
56. As indigenous scholar Dr. Tyson Yunkaporta reports in *Sand Talk* [2019, p223].
57. [Davidson and Noble 1993].
58. [Dixon 2019, p45].
59. [Wilson 1998, p172].
60. [Hiscock et al. 2016].
61. [Kelly 2016, pp49–50], [Lawlor 1991, p157], [Yunkaporta 2019, p152].
62. [Clarke 2003, p75], [Dixon 2019, p39].
63. This supposed hierarchy in technology advancement is supported for instance by Jared Diamond in *Guns Germs and Steel* [1997, p312] discussed further in Chapter 4.
64. [Enders 2015, p226].

65. [Wrangham 2009, pp83–87].
66. [ibid, pp99–102].
67. [Organ et al. 2011].
68. [Wrangham 2009, pp15–36].
69. [Cregan-Reid 2018, p81].
70. [Mithen 2005, p199].
71. [De Waal 2016].
72. [Aiello and Dunbar 1993]. Dunbar later wrote a book about this, *Gossip, Grooming and the Evolution of Language*, published in 1996.
73. [Mithen 2005, pp134–136].
74. [Cane 2013, p183], [Hackel 1985].
75. [Myers 1986, p99].
76. [Mithen 2005, pp114–145].
77. [2005, p146].
78. [2005, p146].
79. [2005, p147].
80. [2005, p150].
81. Steven Pinker in *How the Mind Works* [1997, pp537–538].
82. Henkjan Honing, *Aap Slaat Maat – op zoek naar de oorsprong van muzikaliteit bij mens en dier* [2018] (which means something like: the monkey has the beat – in search of the origins of musicality in humans and animals). The book is translated in English as *The Evolving Animal Orchestra – in search of what makes us musical*, published by MIT Press in 2019). See also the journal article 'Without It No Music – beat induction as a fundamental musical trait' [Honing 2012].
83. J. J. Gibson developed these insights and presented them in his books *The Senses Considered as Perceptual Systems* [1966] and particularly in *The Ecological Approach to Visual Perception* [1979], discussed in detail in Chapter 9 *Experiencing*.
84. The beat perception of newborn babies is described in a journal paper [Winkler et al. 2009] and Honing's book *Aap Slaat Maat* [Honing 2018, pp30–32].
85. [De Waal 2019, p83].
86. [Bramble and Lieberman 2004], [Mithen 2005, pp153–154].
87. *Primate Change – how the world we made is remaking us* [Cregan-Reid 2018].
88. [Steward et al. 2020].
89. See for an overview Josephine Flood's summary of the research in *The Archaeology of Dreamtime* [2010, pp27–38], and also archaeologist Scott Cane in the book *First Footprints – the epic story of the first Australians* [2013, pp17–26]. Cane's book is related to the four-part ABC documentary by Bentley Dean and Martin Butler in 2013.
90. [Mithen 2005, p249], [Pinker 1994, p40].
91. Daniel Everett, *Language – the cultural tool* [2012].
92. [Wilson 1998, pp170–171].
93. Dutch historian Rutger Bregman discusses this in the case of Altamira, in his book *De Geschiedenis van de Vooruitgang* ('the history of progress', in Dutch) [2013, pp51–52].

94. [Mithen 2005].
95. Petzinger's research is presented in a popular TEDx talk (August 2015) and in the book *The First Signs – unlocking the mysteries of the world's oldest symbols* [2016].
96. [Hoffman et al. 2018]. Genevieve von Petzinger, characteristically in a 'vlog' (video blog), posted an excited response to this find.
97. [Higham et al. 2014].
98. [Ackerman et al. 2016], [Wong 2018].
99. Science editor Hendrik Spiering of the Dutch newspaper *NRC* presented these findings in an article on 1 October 2020 (and in the *NRC* weekly edition of 12 October). The research carried out by Hugo Zeberg and Svante Pääbo of the Max Planck Institute for Evolutionary Anthropology, Leipzig, Germany was published online on 30 September 2020 in the science magazine *Nature* [Zeberg and Pääbo 2020].
100. As Hendrik Spiering discusses in an article in June 2020 in the *NRC*, 'Sapiens komt uit een Afrikaanse wolk' ('Sapiens emerged from an African cloud'). The headline on the front page of the science section of the paper asked "Van wie stammen we nu weer af?" (something like 'Who is it this time that we are descendants of?'), emphasising the ever changing and developing insights into the details of our descent [Spiering 2020].
101. Steve Jones, *Darwin's Island* [2009, p2].
102. [Sheldrake 2020, p82].
103. As Merlin Sheldrake discusses in the chapter *The Intimacy of Strangers* [2020], see also Tor Nørretranders's discussion in *The User Illusion* [1998, p337].
104. As Jon Turney discusses in *I, Superorgamism – learning to love your inner ecosystem* [2015], and see also Julia Enders's book *Gut – the inside story of our body's most under-rated organ* [2015].
105. [Lovelock 2019], [Flannery 2010, pp32-39], [Nørretranders 1998, p333].
106. Presented in an article in *Nature* [Simard 1997], introduced by an article by mycorrhizal biologist David Read with the title 'The Ties that Bind' [1997]. The term wood-wide web was on the cover of the magazine [Sheldrake 2020, p169]. German author and forester Peter Wohlleben wrote a bestselling book about the connections between trees, translated in English in 2016, *The Hidden Life of Trees*, and Suzanne Simard published a popular science book on the topic, *The Mother Tree* in 2021.

Chapter 4

Settling

Agriculture, mechanical tools, and writing

Coinciding with the end of the last Ice Age, from about 12,000 years ago, was the start of the Neolithic (the 'new' and final part of the Stone Age) period, during the Holocene geological epoch. From this time on there is a substantial change from hunter-gatherer nomadism to a more settled life in certain areas of the world. Although previously in places with an abundance of food, occasionally some form of settlement in villages emerged; until the settling period people generally moved around in little groups of a few tens of people. But it is important to see this as a gradual transition; semi-settled life did occur depending on climate, influence of seasons, and as mentioned opportunities for food supply. Relationships between tribes' regulation of territorial aspects were important too. At the end of this chapter a reflection on the virtues and losses associated with this transition is presented.

In the fully settled societies a number of interrelated developments took place, such as the development of agriculture, domestication of animals, the invention of writing systems (from mostly pictorial to fully abstract), creation of number systems and mathematics, the emergence of cities and hierarchical social structures, and a number of technological developments, all of which are covered in this chapter.

AGRICULTURE AND DOMESTICATION OF ANIMALS, FORMING OF SOCIETIES

The development of agriculture may have been driven by the fact that the hunter-gatherer way of life was leading to resource depletion, relating to increased population numbers.[1] Using their knowledge of plants and seeds allowed people to increase the yield of crops by planting them deliberately. The invention of agriculture and the domestication of animals allowed groups of people to settle in one location, form larger groups, create villages and eventually cities, and invent writing systems.

This emerged in several locations around the globe, sometimes independently, but first and most prominently in an area this is currently known as the Middle East. This area, often called the Fertile Crescent, is the

DOI: 10.1201/9781315373386-4

area spanning in an arc shape (hence the name) from Egypt around the river Nile in the west, along the east coast of the Mediterranean Sea with Phoenicia, and stretching southeast with Mesopotamia along the rivers of the Euphrates and the Tigris towards the Persian Gulf. As Jared Diamond points out in his book *Guns, Germs and Steel – a short history of everybody in the last 13,000 years* [1997], this is not because the people in this area were necessarily cleverer than others (as has been thought for a long time), but because the climatic and biological circumstances were crucially optimal for these practices to emerge. Not many plants are actually suitable for agriculture, and as it turns out many of those that are suitable were already available in the Fertile Crescent, making it easy for the inhabitants in that area to cultivate these plants. Similarly, of the limited number of animals that are suitable for domestication, many were already present in the area, such as sheep, goats, pigs, cows, and donkeys.[2] Places where agriculture and domestication of animals arose much later and independently, such as the continents of the Americas, had less access to suitable plants and animals.

This was a major development in human history, and indeed for the world – for better or for worse, as will be discussed later. Natural selection (survival and development of species under natural circumstances) was surpassed by artificial selection (driven by variation under domestication) which allowed biological evolution to develop at a much faster pace. Charles Darwin dedicated the whole first chapter on these processes of variation under domestication in *The Origin of Species* [1859] from his own research as an active gardener (he described his profession as 'farmer' at the time). Darwin carried out extensive experiments on his crops to study selection processes, also reported in more detail in his later books. Although his revolutionary insights were initially inspired by his famous five year journey on the Beagle from 1831 to 1836 (like Alfred Russell Wallace, who independently conceived the evolution theory, also after extensive fieldwork and travels to remote places), Darwin's ideas were further developed and refined by his decades-long work on plants at home in Down House in Kent in England rather than just observing the finches on Galapagos Islands. This is the topic of the book *Darwin's Island – the Galapagos in the garden of England* by geneticist Steve Jones [2009], as mentioned in Chapter 2 *Evolving*.

Structure of societies

For the development from migratory, nomadic (hunter-gatherers) to settled ways of life, it is useful to discern several stages related to group size. Throughout his book *Guns, Germs and Steel*, Diamond uses the distinction of *band*, *tribe*, *chiefdom*, and *state*. Though Diamond indicates that some anthropologists describe more categories, this categorisation seems appropriate for the purpose of understanding interaction.[3]

The smallest group, the *band* or clan, would usually be an extended family of a few tens of people, and typical for nomadic hunter-gatherers. Slightly larger is the *tribe*, who in addition to kinship connections share other common factors such as territory and habits, and who have some form of hierarchy but are still quite egalitarian. A tribe can be as big as a few hundred people, still small enough for everyone to know everyone and make decisions based on consensus. The people in a tribe are usually settled in villages, or herding livestock. The next size of group is the *chiefdom*, which has typically a few thousand people, and therefore depends on a more hierarchical power structure and some form of bureaucracy (including rudimentary taxation) and specialisation among its members. This type of society started to emerge from the beginning of settlement, based on the advances of agriculture and domestication of animals. The largest (and currently the prevailing) society is the *state*, with a centralised government, an elected or hereditary leader, bureaucratic structure, taxation, etc.

The transition from nomadic hunter-gatherers to settled farmers and citizens is largely about the development of bands and tribes to chiefdoms and states. In this new situation, establishing a sedentary life in villages and eventually cities, in many cases a number of subsequent changes occurred. First of all, due to the availability of food surpluses in some instances, a much more hierarchical structure of society emerged, and specialisation became possible with dedicated bureaucrats, warriors, and craftspeople. This inherently led to a detachment of maker and user of a tool or technology; previously most people would make and design their own tools.

Although the bigger groups of hunter-gatherers would sometimes have a 'strongman' or leader, this was rare, and only in the sedentary societies chiefs and kings started to emerge and become more prominent. Their power was gained by having control over the goods produced, rather than merit or knowledge. The leaders would ascertain their power through a structure of specialised bureaucrats to manage the storage and trade of goods. The potential for warfare was taken to a new level, as dedicated warriors were used by the powerful leader to conquer new territories, with the aim of gaining more power.

Writing and tools

Artefacts from this time onwards show an increased sophistication, both in functionality as well as in symbolic significance (art and representation, through painting and sculpture) due to the time spent on these developments by dedicated craftspeople. The emergence of hierarchical structure and specialisation played a role in the development of a new range of tools including weapons.

This hierarchical society also accelerated the invention of writing techniques. It is often noted that there is an irony here; the first attempts of writing seem to have been for the purpose of bookkeeping and trade. Rather than being driven by the desire for poetic expression, the nature of the first written records was of an administrative kind (similarly, from the mid 20th century onwards, computers and interactive systems have been primarily used for bureaucratic purposes). This is however not so strange if one realises that at that point oral culture seemed sufficient for this purpose, and can still be considered as a very sophisticated ability to relay stories, contain and transmit knowledge, and develop insight. Only later the new tool evolved into written languages and subsequent cultural changes. This is in analogy with biological evolution, where 'inventions' are stacked and slowly incrementally develop into complex structures, often with intermediate steps in a different direction – the bird's wings for instance (feathers first developed for thermal reasons, and where later in the evolutionary processes adapted through exaptation for flight). Cultural development is quite similar to biological evolution, in that aspect of chance adaptations turning out successful for different purposes, and also often in multiple stages.

Writing developed from ideograms which mimetically depict objects, actions, and concepts, to more abstract and symbolic forms such as text, as is discussed in more detail in Chapter 6 *Communicating*, the language and representation chapter. For the hierarchical and bureaucratic power structures to work successfully, writing was essential. Through writing it became possible to carry orders and information over distances, enabling larger communities to function, and as such an essential element in the development of larger societies.

The technique of writing was invented independently in a few distant areas of the world, in Mesopotamia (3000 BC), Mexico (600 BC), and China (1300 BC) (although the latter might have been connected to the use of writing in Mesapotamia). This is remarkable, due to the difficulty of making such an invention. All other places where writing emerged did so as a result of the influence of cultures that had already invented writing, borrowing and adapting the ideas. Jared Diamond uses the terms 'blueprint copying', for situations where cultures have access to the details of the technique (writing, in this case) and can copy the idea with some modifications, and 'idea diffusion' which is the case when cultures are more distant, and with some reverse engineering re-invent some of the details.[4] Copying and diffusion are points along a continuum of course – any form of inspired invention is possible. It is also worth noting that not in all societies did the settlement and development of hierarchical structures lead to the invention of writing systems.

Ceramics

As stated before, settlement brought about an increase in the invention of tools and use of materials. Previously, the nomadic way of life prohibited the possession of many objects, as it is too cumbersome to carry a lot of things around. Ceramic structures are examples in the category of **chemical technology**, the structure of the material changes with the application of heat: sun dried in the most primitive form, and more advanced with the use of fire, particularly in ovens or kilns. Although pottery was invented possibly as early as 30,000 years ago, for making vessels for transporting liquids and for making ceramic objects such as sculptures, the technique required a settlement due to the time required for the process. This practice furthermore relied on the presence of the right type of clay, which was available around big rivers in Europe, China, Egypt, and Mesopotamia.

The oldest archaeological finds are clay sculptures in Europe from around 31,000 to 22,000 years ago. These have highly stylised elements, showing that rather than just mimetic objects they would have had a symbolic function as well. These objects would have played ceremonial and communicative roles, as well as serving as memory aids in oral cultures, as Lynne Kelly discusses in *The Memory Code* [2016] and *Memory Craft* [2019]. The oldest parts of ceramic vessels were found in China, from around 20,000 years ago. The earliest forms of writing (hieroglyphs and cuneiform) were made by imprinting or incising on wet clay, before drying or firing it.

Mechanical tools and technology

The development of mechanical structures and machines didn't happen in a linear way. In this section some of the interrelated developments are discussed, as they led to the machines we currently interact with, and involve the structure, the application of metal parts, and the inclusion of, an independent power source.

Mechanical structures

Presumably the oldest mechanical invention was that of the lever, which allows the transformation of a low force (input force) and a long movement into a strong force (output force) with a small movement (mechanical advantage). The megalithic structures – for instance, menhirs, and polypods such as dolmens, from around 3000 to 4000 BC – suggest that levers were used to accomplish the movement of the stones of several tons in weight, and we can assume that the ancient Egyptians used the principle in the building of the pyramids.

Although it seems hard to imagine that wooden sticks and logs weren't used to transport heavy objects before, evidence of the invention of the wheel as such (in combination with an axle) start to turn up between 3000 and 4000 BC in Europe and the Middle East. They were used for transportation, and as a tool in ceramic techniques such as the potter's wheel. These and other mechanical structures allow the change of forces (mechanical advantage), through the application of cogs and threaded axles. In engineering literature it is common to distinguish five 'simple machines' or basic types of machines, based on: the wheel (and axle), the lever, the wedge, the pulley, and the screw.[5]

This is the category of **passive mechanical technology** – structures that use the transmission and conversion of movements, driven by the person operating it – as opposed to **active mechanical technology** which involves an independent power source such as an engine, which is discussed in Chapter 5 *Industrialising*.

Metallurgy

In this same period, mechanical technology changed due to the introduction of metals, starting with softer metals such as gold, silver, and bronze. These metals are soft enough to be worked on and hammered into shape even without heat. Their relatively low melting point made it easier to achieve liquid form (for casting) through the use of fire and kilns, a practice explored through the development of pottery in previous times. (This is a good example of how one technology is used to develop the next.) This experience eventually led to the ability to make tools out of iron (and later, steel) which of course led to more effective weapons but also to more intricate mechanical tools for a wide range of useful purposes. Metal was often used in combination with wood, to create mechanical structures.

These mechanical machines needed to have interfaces, such as levers and handles. This is a development from the tool object, which *is* the interface as the same time as being the object. A mechanical structure needs specific parts, dedicated for the function of supporting the interaction, and these parts need to be deliberately designed – in an interface.

Power

A further development which was important for the way we interact with technology is the addition of an external source of power, such as animals or natural sources such as wind and other forms of kinetic energy such as flowing water (rivers, streams, tides). It can also be used to conserve energy, by storing water in some form of container (like a water tower), the kinetic potential energy is available at a later stage.

The domestication of animals wasn't just for the purpose of food (or company), but animals were also used for transportation and as power sources for mechanical structures. Previously, all our tools were powered by human means (including slaves, presumably). This use of an independent power source is an important change in the way humans and technology interact, as from this moment on the interaction has three elements: the human, the technology, and the external force driving it. Human-animal interfaces have been developed for this purpose, ranging from saddles, stirrups, and reigns for horses and donkeys, yokes for oxen, leashes for dogs, etc.

Natural sources of energy were used too, as mentioned. Water was used as a power source, with the water wheel, as well as wind (first for sailing, later also wind mills), driving mechanical structures. This is important for the interface, as this has to facilitate the interaction with the power source; the interaction between people and technology fundamentally changed in this situation. This is further discussed in Chapter 5 *Industrialising*.

Water was used for irrigation, an important element in agriculture. The mastery of water developed through the application of canals, gates, locks, etc., as part of transportation systems for freight and people.

DISEASES

An important aspect of the sedentary societies was that, due to the increased density of people living together, diseases would develop and spread much faster with often devastating results. This was further accelerated due to the situation of living in close proximity to domesticated animals.[6] The increased occurrence of diseases remained the case for millennia, for instance in mediaeval times, the Black Death and other plague pandemics wiped out substantial parts of the populations in the Western world. In the early 20th century, the 1918 flu killed more people than the First World War did (although there is a connection between the two events, in that by supressing publicity around the disease due to the war – even the name "Spanish Flu" was part of the propaganda as Spain was neutral – the countries at war attempted to deny the origins, which strongly exacerbated the impact of the disease).

The practice of settlement led to many diseases, having a very negative effect on human health, which was a big price to pay. This lasted for millennia, until advances in medical science led to better survival rates – at least in some parts of the world. This was mostly through the awareness of the causes of the diseases (though microbes were discovered in the 17th century, they were not identified as sources of infection until later) and subsequent improvement of hygiene. But even in recent times, outbreaks of various animal-related viral diseases transferred to humans, such as bird flu, swine flu, and most recently the COVID-19 virus spreading around the world since the beginning of 2020, showing that there is still a big risk of epidemics and even pandemics. In fact the risks are increasing, as humanity further encroaches on nature, particularly through practices such as deforestation and wildlife poaching, and further accelerated by international air travel.

Diseases also played an important though largely unintended role in the conquest and colonisation by Europeans of the rest of the world, as Jared Diamond illustrates in his *Guns, Germs and Steel* book [1997]. Colonising forces didn't just have the advantages of the possession of better weapons and access to the accumulated knowledge through written sources, they also carried diseases which they themselves were resistant to. These diseases killed many of the people in the areas they invaded, even if the colonisers didn't intend to (although they might have). This was the case with the colonisation or conquering of Meso-America, for instance. Another example is what happened in Australia after colonisation which started in 1788: diseases that the British settlers carried with them (but were largely immune to) eradicated up to half of the resident population who had no defence against those diseases.

ANCIENT CIVILISATIONS

All the developments in the settlement period led to the establishment of the 'great civilisations' of ancient Greece, the Roman Empire, Mesoamerica, and China, expanding from the cradles of civilisation such as the Fertile Crescent (Mesopotamia, Egypt) with cities and several advanced technologies. Writing made bureaucratic hierarchies and long-distance communication of orders possible, and technology was applied to the development and application of weapons for warfare. Inevitably the scale of wars grew with the scale of civilisation, and often other states that were conquered became a source of wealth in material and people-power (slavery), further increasing the strength of the empires. The classical Greek society developed social and democratic structures, and advanced artistic expression for instance in painting and sculpture. The Roman Empire developed this further, also in technology and infrastructure (roads, aqueducts) and organisational structures. China was usually more advanced in technology than the West, at least until 1400 AD, but isolated from the West, so many inventions were not communicated.

It might seem a bit unfair to say so little about these great civilisations, surely they deserve more attention than this scant coverage here. After all many books are written about the ancient Greek and Roman cultures. But precisely because of the fact that so much attention usually is given to them, and because it seems mainly an accumulation and extreme refinement of earlier developments as covered above, I will instead put more emphasis on the next phase, of industrialisation, the Renaissance, and the Enlightenment. This phase is discussed in the next chapter. Rounding off the coverage of the settlement phase is a reflection on the balance between nomadic and sedentary societies, and the importance of the conservation of knowledge during transition periods.

NOMADS VERSUS SETTLERS – THE VALUE OF TRADITIONAL KNOWLEDGE

It is good to bear in mind that in the transition from nomadic to sedentary lifestyle, even though it led to the civilisation as we know it, we also have lost a few things.

There are several anthropological researchers who are urging us to respect the knowledge of traditional cultures and demonstrate how ancient knowledge is still relevant, for instance, Jared Diamond as mentioned earlier in his *Guns, Germs and Steel* [1997], and more specifically in *The World Until Yesterday – what can we learn from traditional societies?* [2012], Wade Davis in *The Wayfinders – why ancient wisdom matters in the modern world* [2009], and Dan Everett in *Language – the cultural tool* [2012].

Everett (as mentioned in Chapter 2, and in Chapter 6) writes about the Pirahã people in the Amazon, who were isolated from contemporary society until recently. They reject many contemporary technologies and techniques, and for good reasons as Everett emphasises, it is not because they are backwards, but because they don't need them.

Diamond writes particularly about New Guinea (where he worked in ornithology and ecology for years), where until the relatively recent Western colonisation about one million people lived in small groups separated through the particular geographic circumstances, leading to relative isolation (and animosity) between groups – about 1000 of the world's 6000 languages are spoken here.[7] As mentioned before, New Guinea and Australia used to be one continent (Gondwana), and became separated about 10,000 years ago. Probably partially due to the different landscape and a sparser inhabitation, Australian culture developed differently. And although also developing about 250 languages (all different, though related, and even more dialects),[8] the instrument of the Tjukurrpa (or Dreaming, or lore, see below) connected cultures across the continent, diminishing the need for warfare in response to conflict.[9]

Traditional nomadism

Although a somewhat romantic notion, we might consider the benefits and opportunities of the nomadic over the sedentary ways of living, as advocated by authors such as Bruce Chatwin in *The Songlines* [1987] (and in his other books), or Robyn Davidson in *Tracks* [1980], writing about their experiences of learning about Australian Aboriginal culture. Davidson is well known for her exploration of the Australian landscape and culture; starting from Alice Springs, after a long preparation, she walked with her camels 2700 km westwards through the deserts to the coast in 1977. Although unintended, and at times unwanted, due to the coverage in the *National Geographic* and the impressive images of photographer Rick Smolan,[10] her quest became world news and brought many new insights to a mainstream public.* Davidson further explored nomadism by participation in the journeys of the Rabari nomadic people in Rajasthan in northwest India, presented in her book *Desert Places* [1996]. An insightful reflection on these experiences and both the practical and the historical contexts can

---

* *Tracks* has been a bestseller since it came out in 1980, but it wasn't until 2013 that the story was made into a movie. The film, starring Mia Wasikowska and Adam Driver, inevitably leaves out many aspects of the book, although the cinematography of the landscape is impressive (as were the original photographs of Smolan). What come across in the book is that Davidson's journey and preparation were the goal in itself, not just about reaching the end destination, very much like the *dérive* – wandering, but not being aimless. The film in a way tries to explain this unexplainable drive mainly through a more one-dimensional version of the character's determination, leaving out other essential aspects and the internal contemplation. The role of the Pitjantjara elder Mr. Eddie as a voluntary guide and generous source of geographical and cultural knowledge [Davidson 1980, p165], similarly has lost some of the more interesting facets. Robyn Davidson's *Tracks* book is written from an insider's viewpoint, from the author's autobiographical perspective, while the film only shows what it looks like from an outsider's perspective, inevitably more superficial in that aspect.

be found in her essay 'No Fixed Address – nomads and the fate of the planet'.[11] Nomadism nowadays is influenced by recent historical developments; the Rabari travel using domesticated animals (camels). In Davidson's essay about nomadism she suggests that the biblical story of Cain and Abel, as a farmer (settler) and a shepherd (nomad) respectively, could be read as a metaphor for the tension between the nomad and the settler (their rivalry leading to the 'first murder').[12] This point is also made by Bruce Chatwin, for instance, in *The Songlines*.[13]

The gradual transition from nomadic to sedentary life is important in human history, as it includes the start of the transition from oral to literal culture, development and use of new tools (passive mechanical technology), domestication of animals, and cultivation of plants, but it is not necessarily an entirely positive development. We have paid a price. By uncritically following the opportunities offered by the new technologies and inventions, cultures and societies have been influenced and structured in ways that were not always optimal for humankind – and certainly not for other animals, or nature in general. Sometimes it is good to reflect on the historical developments, and if we realise that it is not a linear accumulative progression towards an inevitable goal it is possible to learn how to further improve our current situation. This point is made in the context of human evolution, for a long time seen as a 'ladder' of linear progress; now we know it is a chaotic development of interrelated complexities.

We have the opportunity, and the ability, to view and guide the developments in different ways. The transition from hunter-gatherer to sedentary culture didn't happen in just one way, as a linear story of progress – there have been many different developments, many of which we can learn from.

Degree of settlement in traditional Australian Aboriginal culture

A particularly interesting situation evolved on the Australian continent. As the author and scholar Eric Rolls points out in an insightful response to Davidson's essay 'No Fixed Address' about nomadism, in Australian Aboriginal culture settlement did occur in various degrees, managing the landscape with fire, and constructing agricultural structures.[14] It is now becoming increasingly clear that the idea that the Aboriginal people were just wandering primitive nomads was a narrative perpetuated by the early colonisers in the late 18th and early 19th century in particular. This was a narrative to support their claim that the continent was effectively uninhabited, a 'terra nullius', ready to be exploited and appropriated for their needs – to use as a penal colony, and for agriculture, among other purposes.

Recent insights reveal the full extent of the ancient practice of 'firestick farming' by the original inhabitants of Australia (see below) which effectively cultivated the whole continent, as historian Bill Gammage elaborately describes in his book *The Biggest Estate on Earth – how Aborigines made*

Australia [2011], the biologist Tim Flannery in a chapter in his book *The Future Eaters* [1994, pp217–236], and Victor Steffensen from his own Indigenous perspective in *Fire Country* [2020]. Since the British settlers arrived in 1788 with the "First Fleet" of convicts and military,[15] what they encountered was described as 'parklands'. This is captured in many paintings of that era, as well as in many written accounts which sometimes were thought of as presenting a romanticised image; however we now know this was actually accurate. In his book, Gammage compares these paintings with the current state of the sites, which shows how the Australian landscape since the end of Aboriginal custodianship has degraded to barren, overgrown, or otherwise unhealthy 'bush'. The common view on the Australian outback, the bush as a kind of wilderness, is not how it was under Indigenous custodianship. It was well maintained, they kept the 'bush' healthy to suit their purpose of hunting and enabling certain flora to flourish; there are several plants that only germinate when the seeds are charred by fire, including several edible ones. Their traditional knowledge of fire, flora, and fauna is profound, and was applied in this intricate ecological practice.

Furthermore, it is now increasingly becoming clear that the widespread proof of Indigenous forms of sedentism and settlement were often ignored and even actively suppressed from the beginning of British colonisation. A different picture is currently emerging, showing that the traditional Aboriginal culture had a mix of nomadic and settled forms of living. Evidence for this is persuasively presented in the book by Bruce Pascoe *Dark Emu – Aboriginal Australia and the birth of agriculture* [2018].[16] Pascoe goes back to the reports of the colonising explorers, combined with knowledge from recent academic studies (such as the work by Bill Gammage, Eric Rolls, and others) in Indigenous land use, and brings together a large body of evidence and insights that support the notion of a sophisticated society with several degrees of sedentism. As Pascoe summarises "Aboriginal people *did* build houses, *did* build dams, *did* sow, irrigate and till the land, *did* alter the course of rivers, *did* sew their clothes, and *did* construct a pan-continental government that generated peace and prosperity."[17] Traces of extensive fish farming and traps using stone walls can still be found in the landscape, including remains of houses and traces of eel smoking practices[18] such as in the Budj Bim cultural landscape of at least 6,600 years old at Lake Condah in Victoria (it was added to the UNESCO World Heritage list in July 2019). Grindstones were used to prepare seeds for making bread, possibly as early as 30,000 years ago.[19] Although Pascoe has been criticised for exaggerating his claims and inconsistencies in his scholarly research[20], the immense popularity of *Dark Emu* plays an important role in advocating an understanding and appreciation of Indigenous culture and knowledge, which was severely lacking in even the recent past. Nowadays there is even a children's book *Young Dark Emu* used in school curricula. Several friends and colleagues have told me they didn't learn about this in school, and even Lynne Kelly (author of the books about 'memory spaces') writes in *Songlines* "Why, oh why was I taught nothing at school about Aboriginal intellectual achievements?"[21]

We can see that the development of agriculture in general went through stages, similar to the four stages of tool development as presented in Chapter 3. One could identify the 'opportunistic land use' (similar to the first tool stage of found objects appropriated for a certain task) as a hunter-gatherer approach, and the 'dedicated land use' where the land was actively manipulated to serve a certain purpose (early forms of agriculture and domestication of animals), then the planned (designed) use of land, and composite land use. Like with the tool development stages, these are all interrelated and not necessarily hierarchically organised, All of these stages were, in more or lesser extent, in use in traditional Australian Aboriginal culture. But most prominently in their culture there are strong relationships between spiritual, law, and practical knowledge.

Dreaming and songlines

Beyond the mere caching of surplus food resources, or stockpiling for major events, food was apparently stored in large quantities, in storage chambers made of clay and straw.[22] Ironically, as Pascoe points out, several of the early colonial explorers used these food sources for their survival.[23] Traditionally locks and fences, for example, weren't necessary in a culture based on consensus rather than combat – Pascoe calls it "jigsaw mutualism" – each piece governing the rights and responsibilities locally while maintaining an insight in the whole puzzle.[24] Ownership rights and responsibilities are regulated and maintained through the laws or Tjukurrpa (Dreaming), the holistic knowledge system, which also includes the relationships between the people and the land, stories, song, and dance. This concept is a bit hard to grasp for the Westerner's mind set (such as mine); the notion of Tjukurrpa is about religion ('Dreaming'), knowledge, law, social rules and morals, and history, as the archaeologist Scott Cane explains in the book *First Footprints* of 2013 based on accounts of the Elders he worked with.[25] The term 'Dreaming' is, as Indigenous scholar Tyson Yunkaporta in his book *Sand Talk* [2019] states, "a mistranslation and misinterpretation", but he explains that like with many English words that are used to refer to Indigenous knowledge concepts, it is just a label, a shorthand, however inadequate. Yunkaporta proposes as an attempt for a full description in English of the Dreaming concept – "supra-rational interdimensional ontology endogenous to custodial ritual complexes" – quite a mouthful indeed. "So Dreaming it is", he concludes.[26] Bruce Pascoe explains Dreaming in *The Little Red Yellow Black Book – an introduction to Aboriginal Australia* [2018] as part law – "the rules for living, languages, customs and ceremonies" which go back to the creation stories of the ancestral beings. For this then, "we use the English word 'Law' to mean this entire cultural inheritance, its injunctions and taboos."[27]

Tjukurrpa or Dreamings are communicated over the vast distances across the whole continent through the intangible network of 'songlines'.[28] Bruce Chatwin wrote his book *The Songlines* in the late 1980s, bringing broader attention to the concept of Songlines and other aspects of traditional Australian Aboriginal culture, which was not widely known at the time. Although the book has its limitations, which Chatwin admits - as it was not easy for a white person (and a 'pom' – Australian slang for British visitors) to get access to the traditional knowledge which he greatly respects, he did try to draw attention to the many misunderstanding and tensions between the cultures. In the book there is a good description of the Songlines principle as told to him by an Aboriginal Elder,[29] capturing some of the essence, though more of the complexity is covered in the other, more recent, and more authentic sources mentioned in this section, such as the recent book *Songlines – the power and promise* by Margo Neale and Lynne Kelly [2020]. Emphasising the scope of the concept, Indigenous scholar Neale (Head of the Indigenous Knowledges Curatorial Centre at the National Museum of Australia) writes "Songlines are foundational to our being – to what we know, how we know it and when we know it. They are our knowledge system, our library, our archive from which all subjects are derived."[30] These Songlines can stretch for hundreds, even thousands of kilometres. A narrative can travel along these vast distances, changing language where appropriate, and for instance used for trade. The Songlines, or Dreaming tracks, link locations to knowledge, through rituals (including song and dance), as a way of memorising and transmitting knowledge. This principle of memorising knowledge through locations is a well-known technique particularly in non-literal cultures, turning the landscape into a "memory space" as Lynne Kelly discusses in the book *The Memory Code* [2016]. After studying how oral cultures relate knowledge to physical items and landscapes, the "encyclopaedic memories of the Elders", particularly with Australian Indigenous knowledge keepers, Kelly presents many memory spaces around the world that she studied as memory spaces. These include sites such as Stonehenge, Easter Island, megaliths in Carnac, the giant sand drawings at Nasca, and the architecture of the Pueblo people. The common interpretation is that these structures were for ritual and religious purposes, which doesn't say very much. Of course, it cannot be proven, but if one tries to understand knowledge structure in oral culture, it makes sense that these structures were memory spaces, just like the Songlines. The giant glyphs on the desert floor at Nasca for instance are generally constructed from one uninterrupted and unambiguous line which can be walked (rather than being made by or for aliens, as some have tried to explain these structures in the past,[31] driven by ignorance and deep underappreciation of traditional knowledge) (crop circles are still very popular too[32]).

From an engineering perspective it is interesting to note that through the Songlines, knowledge travels in a similar way to how data travels through a 'mesh network', where each node is able to receive and propagate the messages (as opposed to a centralised network). A further interesting analogy with the notion of decentralised networks in contemporary engineering practice is that the traditional Aboriginal society is non-hierarchical and distributed, just like the Internet which has no central 'server'. In biology, there is a strong analogy in the mycorrhizal networks that link roots of plants and trees with fungal hyphae, or the distributed nervous system of the octopus.[33]

In traditional Aboriginal culture there is no 'king' or strongman, but instead there are Elders who have to earn their recognition based on merit (such as knowledge possessed), not power or material possessions.[34]

The Songlines are an interconnected network of cultural and spiritual knowledge of the Dreaming or the Tjukurrpa, that resonates across the whole continent.

Firestick farming in traditional Australian Aboriginal culture

It is particularly interesting to look at the use of fire in traditional Australian Aboriginal culture. As mentioned above, fire is used in order to change the landscape and to maintain it, effectively using fire as a farming or cultivating technique. The ecosystem of Australia, as a rather hot and arid continent, has included bush fire as an essential element for millions of years, and in many ways even depends on it.[35] The Australian Aboriginal people are extremely proficient in their knowledge and application of fire, using skills and insights developed ever since they first arrived on the continent, effectively managing the land with fire. Their technique keeps the fires under control. Uncontrolled fires that usually make the news completely destroy bushlands and vast areas of forests, and often built-up areas. For example, one of the worst events was on 7 February 2009 ('Black Saturday') when 173 people died in bushfires north of Melbourne, caused by a combination of natural events, electricity line malfunctions, and arson. The recent fire season on the south-east coast of Australia was longer and more intense than ever before, known as the 'Black Summer' stretching from early October 2019 to the end of January 2020, resulting in 33 deaths, thousands of homes destroyed, several million hectares of land burned – including rainforests and wetlands (!), many sacred Indigenous sites lost, and possibly more than a billion animals killed. In the last months of 2019 the bushfires around Sydney were so prominent that the city was covered in smoke, with orange sunsets starting mid-afternoon,* and being a severe

---

* reminiscent of the key scenes in *Blade Runner 2049* from 2017, in the post-apocalyptic and radio-active wastelands where Decker (Harrison Ford's character from the original Blade Runner movie from 1982) hides out – apparently this was inspired by (photos of) the dust storm that turned Sydney orange in 2009.

health hazard. Driving to Melbourne in mid-December 2019 had a very post-apocalyptic feel, we were surrounded by an orange haze until way past the Victorian border.

Among the many advantages of the privilege of living in a country with a still surviving Indigenous culture of over 50,000 years old, are the practical lessons that can be learned from experiencing, studying, and learning about their knowledge of tool making, mastery of fire, and their very sophisticated culture based on a deep understanding of nature and technology. This is a culture where nature, technology, and culture are resonating together.

In July 2012 I participated in an Indigenous bush fire management camp in Cape York, for a project by my colleague Jacqueline Gothe (Associate Professor in visual communication design at UTS) who has had ongoing collaborations with Aboriginal people and Indigenous-led research projects for decades, particularly with Victor Steffensen (regular UTS collaborator) who for many years has worked on capturing traditional knowledge using video and new media.[36] I went to participate in the workshop with a videographer colleague, to capture image material and soundscapes of the bushfires and learn about this practice, camping and exploring for several days. The participants of the workshop were mainly rangers and others professionally involved in forestry and contemporary bush care, including Indigenous people.[37]

Observing the fire practices in several sessions, I learned that rather than ravaging bush fires that make the news, the Indigenous technique applies what is called 'cool burning'. The fire travels fast through the woods and only burns the low scrubs (undergrowth), unlike a hot fire which would burn the whole tree (the canopy is sacred in Aboriginal culture, and as always there is a reason for it), as Victor Steffensen, one of the workshop leaders, explained during the workshop and in more detail in his book that was published at the beginning of 2020 (well timed, right after all the four months of intense fires on the southeast coast of Australia), *Fire Country – how Indigenous fire management could help save Australia*.[38] During the workshop, Victor often walked barefoot over surfaces that they had just burned, unintentionally emphasising this 'cool burning' (though I struggled a bit to do the same, the soles of my feet are a bit too sensitive).

Indigenous bush fire management is a profound example of a holistic, ecological approach, as they skilfully take into consideration a large range of interrelated environmental factors. Here is the list I made (which I checked later with Victor): type of plants, time of year, air temperature, moisture levels of the vegetation, animals present, insects (many climb up in the sprigs and trees to survive), birds attracted to the insects, birds of prey, wind speed, wind direction, humidity of air, type of soil, and sun light levels (there are more examples in Victor's book). To give an example of the latter, one day the workshop participants burned in one area, the next day when it was overcast we went to a different part of the woods. This is very different from the Western approach, as Victor emphasises, which, although the technique of preventive burning ('backburning') is now common practice, it is generally quite meticulously planned ahead and doesn't have the flexibility to respond to the dynamic local environmental factors.

There have even been eyewitness accounts of particular birds (raptors such as kites and falcons) in Northern Australia practising the same technique – they pick up burning embers and drop them elsewhere, controlling where the burns are as a hunting technique. These events are being studied from a scientific perspective, but the practice has been widely known in traditional Indigenous culture.[39] And like some other animal tool use, we can't actually be certain that these practices are recent (copied from humans) or actually preceding human application.

Bruce Pascoe in *Dark Emu* describes five principles of the Aboriginal approach to fire: a "rotating mosaic" of firing of land areas (controlling intensity, and allowing some plants and animals to survive), taking into account the time of year, the weather, informing neighbouring clans of the fire activity, and finally, avoiding the growing season of particular plants, avoiding potential food burning.[40] The fear of bush fire has been understandably very common for Australians who arrived after 1788, and it was therefore quite an experience to go into the woods with a group of people (as mentioned, many of the other attendants to the workshop were bush rangers other people professionally involved in forestry), and watch the workshop leaders (and later the participants too) set fire to the bush, explaining how they did it, and correctly predicting where the fire was moving and for how long. In some cases the whole group just drove off, leaving the fire to burn through the landscape. They would return later to check that the fire had stopped exactly where they predicted it, which showed to me that they really understood and mastered the fire ecology. The power, and the importance, of this type of preventive burning if done properly, as an essential 'maintenance' of the landscape, was made very clear in this event.

Another impressive experience was to take part in a smoking ceremony, a common practice in Aboriginal cultural events. Particular leaves are used, the smoke of which symbolises a cleansing effect.

In December 2012 our team at UTS presented some of the material in an interactive multi-screen audiovisual setting, as a demo working towards an exhibition that the National Museum of Australia was interested in. The demo presented the different components and interactive presentation ideas, such as the real-time manipulation of projected flames through gestures.[41] Although at the time it didn't lead to an exhibition, conversations have continued since, showing that the interest is still there.

Polynesian navigation skills

As mentioned earlier, another publication that presents an argument for the appreciation of traditional knowledge is *The Wayfinders – why ancient wisdom matters in the modern world* by the anthropologist Wade Davis [2009].

The 'wayfinders' in Davis's book are the Polynesians, who are spread over more than 1000 islands in the Pacific Ocean, in an area larger in size than most continents. Davis dedicates a whole chapter to them, explaining how they approached naval navigation covering the vast area of islands in the ocean.[42]

They even navigated to Easter Island, which is at a distance of 2,500 km from the nearest island! And even on this remote and detached location, their culture thrived – in contrast to the common story that uses Easter Island as an example of environmental decline, "an ecological disaster unfolding in complete isolation" as Jared Diamond puts it in his otherwise insightful book *Collapse*.[43] They didn't have it easy, but there was certainly no disastrous decline in resources and widespread animosity between the inhabitants until outside visitors started to arrive. In *Collapse*, Diamond used Easter Island as the key example of human-caused environmental decline, a parable almost of the inevitable collapse that human global society is heading for if we carry on affecting the environment in the way we do. This focus on Easter Island turns out to be not the best choice, of Diamond's insightful and important message. As Rutger Bregman discusses in his book *Humankind – a hopeful history* (based on the recent research of Jan Boersema), the Polynesian inhabitants of Easter Island were actually very healthy and living under relatively good conditions in the society they created in this environment.[44] Diamond's point is still valid of course, we can learn a lot from the history of past mistakes of humankind, affecting the environment and leading to collapses of societies.

Polynesian navigation is a holistic approach, taking into consideration a vast range of cues combined with a continuous awareness of the location of the vessel. This is one of the forms of 'dead reckoning', which is a technique also used with some modern electronic instruments: by keeping track of all changes that occur over time from a known starting position, the current position can be inferred. While electronic systems can be always on, the human navigator using this technique can't afford to sleep. Errors in the pickup of information accumulate with this technique, and over time the estimated position will be further and further off compared to the actual location. However, unlike many electronic systems, the human navigator can take in other cues and adjust the knowledge of the position – this is the strength of the holistic nature of the traditional knowledge and insights. The cues the Polynesian navigators use include the patterns of swells (unlike coastal wave patterns which are essentially chaotic, ocean swells have specific patterns around islands), presence of birds (each species has their own range, and therefore indicate the distance to a landmass), fish and other sea animals, floating debris and flotsam, seaweed, clouds, winds, smells, light patterns (also related to weather, such as a red sky sunrise or a moon halo), the sun, the moon, and most of all, the stars. Navigation by stars however is only possible at night, so the navigator has to remember where the stars are in relation to the horizon when they disappear at dawn, until they reappear at dusk. An expert Polynesian navigator knows the name and position of about 220 stars in the night sky. These navigation and sea faring skills were developed over the millennia, when people colonised the islands from mainland Asia in the west.

Unfortunately, because of the fact that the prevailing trade winds blow from the east, and some other cues, during the 20th century a (false) theory that Polynesia was settled from the east (from South America in pre-Columbian times) had become popular. This was most famously supported by the charismatic Thor Heyerdahl from Norway, who, in order to attempt to prove his misguided theory, in 1947 managed to sail (or drift, rather) a balsawood raft with a crew of five other untrained people from Peru to Polynesia. The book of the journey became a bestseller, however. Several further raft expeditions were undertaken in the years since. One of those was *Las Balsas*, consisting of three rafts led by Vital Alsar from Spain, which managed to reach the Australian east coast (about 750 km north of Sydney). One of the rafts, together with a range of associated paraphernalia, is on display in a purpose-built maritime museum in the landing place Ballina which I visited in 2013. It is an impressive experience to see the raft up close.

Due to the huge success of Heyerdahl's book at the time (and a museum in Oslo dedicated to the trip), the theory is still popular, though history has firmly proven that it is wrong.* In addition to the already known evidence of linguistic and ethnographic nature, recent archaeological and genetic evidence show that the islands were colonised from the west.[45] This was possible through the extreme navigation and seafaring skills and knowledge of the Polynesian culture. Also we now know (and the Polynesians did for much longer), the trade winds reverse at a certain time each year, which would have facilitated some of the longer journeys in a purposeful (not accidental) way. In *The Wayfinders* Wade Davis makes quite a point of this (and this is the reason for me to cover it here), because, as he puts it, "Heyerdahl's theory, which denied the culture its greatest accomplishment, was the ultimate insult."[46] Heyerdahl also tried to apply this misconception to Easter Island, even insisted on a link between the giant statues there (the 'moai') and stone buildings of the Inca culture, and even the pyramids in Egypt. But these and other misguided 'suggestions' such as alien origins are now widely debunked, for instance by Jan Boersema and Jared Diamond.[47]

Although it was quite a feat of Heyerdahl and his crew in 1947 to get the raft across the Pacific, all that he actually proved is that an accidental connection would have been possible, and not more than that. According

---

* A movie from 2012 about Heyerdahl's adventure with the Kon Tiki, though very beautiful and engaging, makes no mention of the fact that his theory was misguided (it isn't very factual on a number of other details either, it was shot around Malta which gives a leisurely Mediterranean feel to the movie, and a scene about a Polynesian island is shot in Thailand with local actors as Indigenous Polynesians).

to Wade Davis it did have one positive effect, in that this incident caused a revival of their ancient naval knowledge. To prove Heyerdahl wrong, it led to the construction and launch of a large ocean-going double-hulled sailing canoe based on a traditional design in Hawaii in 1975, the Hokule'a. It has since sailed across Polynesia, navigating to many of the islands, to other continents, and eventually even around the world, only using traditional navigation knowledge and skills (no instruments).

The value of this navigational knowledge was already acknowledged in the late 18th century by Captain Cook, himself an able navigator even with the limited instruments of his time. On his first voyage starting in 1768, (then Lieutenant) Cook didn't have the advantage of the chronometer and had to work out the longitude by charts, position of sun, moon, and stars, and sighting of land (most sea explorations until the widespread use of the marine chronometer took place by following coast lines by necessity). On his second voyage, when he sailed his ship the Endeavour from Tahiti to New Zealand, Cook was helped by a Polynesian navigator named Tupaia. After this 'discovery' they sailed on, to map out the east coast of Australia in 1770. The Endeavour made first landfall in Botany Bay, and Cook claimed the area for Britain with the aforementioned far-reaching consequences.

Contemporary nomadic tendencies

The urge for nomadism still exists in contemporary Western human culture; we backpack, travel, and hike around the world – though this is usually very safe and only acceptable to us because we know that in the end we can always return home.

As Lewis Mumford put it in 1961 in his book *The City in History*, "human life swings between two poles, movement and settlement."[48] He relates these 'poles' to the animal and the plant kingdom, respectively. We can however find examples between these poles, such as the sessile habit of the oyster from one extreme, to the detachment of seeds of plants from the other extreme (Mumford's examples), the further extreme of the latter as exemplified with the behaviour of runaway tumbleweed.

The Situationists, an art movement in different locations across northwest Europe in the 1960s, developed the notion of the *Dérive*, as an act of wandering through urban environments such as Paris, seemingly aimlessly but deliberately open to serendipitous encounters. The *Theory of the Dérive* was presented by Situationist Guy Debord in 1956. The artist Constant Nieuwenhuys from the Netherlands, who was associated with the Situationist movement between 1958 and 1960, proposed nomadic cities, spread out across the landscape. Instead of the nuclear city and villages we are familiar with, and the related (sub)urban sprawl, his vision of a 'unitary urbanism' allowed inhabitants to roam from place to place in a nomadic style.[49] Constant built scale models and made paintings to support and develop his ideas, still on display in many contemporary art museums (for instance in The Hague in the Netherlands, in *Het Gemeentemuseum*). Although strongly utopian, and at least somewhat impractical, it is worth reflecting on this work as an interesting alternative for the commuting lifestyle particularly prevalent in the Netherlands.

Currently there is a potential for reviving this vision, if only the house- and room-sharing schemes such as Airbnb would be more focused on the original ideas of the 'sharing economy'. It has led to the emergence of the 'anywheres', who can have the same lifestyle wherever they choose to live. In its current form the promises of the sharing economy are not fulfilled because the companies are focused on profit, as the historian Frank Trentmann points out in his book *Empire of Things*.[50] These schemes actually create more demand, like 'ride sharing' initiatives such as Uber that put more cars on the road, and delivery services that contribute to an increase in traffic too. It also accentuates the presence of a whole subclass of people in the 'gig economy', working below the minimum wage because they have no choice, and having no social security or healthcare support. But there is a potential for innovative and more appropriate distribution of goods and services, through the sharing economy. For this we need the focus to shift from purely capitalist motives to true attempts to share. For instance, 'ride sharing' schemes could take traditional hitch-hiking as an inspiration, rather than taxis.

NOTES

1. Dutch historian Rutger Bregman suggest this in his book *De Geschiedenis van de Vooruitgang* ('the history of progress', in English) [2013, p53].
2. Jared Diamond, *Guns, Germs and Steel – a short history of everybody for the last 13,000 years* [1997, p160]. I read the 2005 edition of Vintage; a later edition has the subtitle *the fates of human societies*.
3. In Diamond's later book, *The World Until Yesterday*, which is about the value of the knowledge of traditional and ancient cultures, he uses the same categories. Diamond attributes these categories to the theories of cultural anthropologist Elman Service in the 1960s [2012, p14].
4. [Diamond 1997, p224].
5. The basic machines are for instance described in *Engineering – a very short introduction*. [Blockley 2012, p23].
6. This is one of the key points of Jared Diamond in *Guns, Germs and Steel* [1997]. See also Bill Bryson in *The Body – a guide for occupants* in the chapter at the end of the book, 'When Things Go Wrong' [2019, p325].
7. [Diamond 1997, p306].
8. Aboriginal languages are discussed in *The Little Red Yellow Black Book* [AIATSIS and Pascoe 2018, p42], and in more detail by linguist Bob Dixon in *Australia's Original Languages – an introduction* [2019].
9. Archaeologist Scott Cane presents this in the book *First Footprints – the epic story of the first Australians* [2013, pp17–26]. Cane's book is related to the four-part ABC documentary by Bentley Dean and Martin Butler in 2013.

10. The main article appeared in the May 1978 issue of *National Geographic*, with Davidson (and one of her camels) on the cover. Davidson and Smolan later published a book with more photographs and excerpts from *Tracks*, *From Alice to Ocean: alone across the outback* in 1992.

11. In the magazine *Quarterly Essay* [Davidson 2006].

12. [Davidson 2006, pp5–9].

13. [Chatwin 1987, pp192–193].

14. In the following issue of *Quarterly Essay* [Rolls 2007].

15. As Robert Hughes describes in *The Fatal Shore* [1986].

16. The 2nd edition of Pascoe's book is from 2018; it was originally published in 2014 as *Dark Emu – black seeds: agriculture of accident*? Bruce Pascoe earlier (co-)wrote *The Little Red Yellow Black Book – an introduction to Indigenous Australia* [AIATSIS and Pascoe 2018] (the 4th edition, the 1st edition was from 1994).

17. [Pascoe 2018, p183].

18. Bruce Pascoe discussed the eel farm structures at length in Chapter 2, 'Aquaculture', in *Dark Emu* [2018, pp68–96], including the evidence of eel smoking practice [2018, p84]. The eel farms are also mentioned by Jared Diamond in *Guns, Germs and Steel* [1997, p310], as well as the cultivation of seeds [1997, p311].

19. [Pascoe 2018, p30].

20. Most prominently in a 2021 book by anthropologist and linguist Peter Sutton and archaeologist Keryn Walshe, *Farmers or Hunter-Gatherers? - the Dark Emu debate*. Pascoe welcomes the debate, it is such an important topic and it needs more research, more books, and more discussion. It wasn't helpful however that the Sydney Morning Herald newspaper announced the new book with the sensationalist headline "Debunking Dark Emu" on the front page of the 12 June 2021 issue, though the actual article by journalist Stuart Rintoul (in the *Good Weekend* section, next to an advertisement for Kimberley Adventures: "wild, rugged yet perfectly civilised") with interviews with the authors Sutton and Walshe, and a response from Pascoe, is more nuanced. Several other articles written in the Guardian, The Conversation and Inside Story present a good balance, but overall the polemic tone of the debate is not helpful to the cause. Sutton and Walshe make some important points, such as that we shouldn't think of a hierarchy, of agriculture being better than hunter/gatherer culture, and that the key aspect is the way the culture relates to the land, which is indeed what I have been reflecting on throughout this chapter.

21. *Songlines – the power and promise*, by Margo Neale and Lynne Kelly [2020, p7].

22. [Pascoe 2018, pp147–148].

23. [Pascoe 2018, p34].

24. [Pascoe 2018, p199].

25. [Cane 2013, p80, p166].

26. Tyson Yunkaporta in *Sand Talk – how Indigenous thinking can save the world* [2019, p22].

27. *The Little Red Yellow Black Book* is assembled and published by AIATSIS, the Australian Institute of Aboriginal and Torres Strait Islander Studies [AIATSIS and Pascoe 2018, pp12–13].

28. [Cane 2013, p80, p166].

29. [Chatwin 1987, pp56–60].

30. [Neale and Kelly 2020, p3].
31. Swiss pseudoscientist and known fraud, Erich von Däniken, has written several bestsellers since the 1960s, for a while making the idea of alien connections in palaeolithic times widespread – including the moai sculptures on Easter Island (see below).
32. As I noticed in two recent editions of 'alternative' magazines, which devoted articles to the phenomenon of crop circles. Even though widely known as human made (the original makers since the 1980s have revealed themselves as such), some people still believe other possibilities. Andy Thomas (who authored several books about conspiracies) in the article 'Crop Circles of 2020 – irrepressible persistence' in *Nexus* ("the alternative news magazine", 27/6, pp48–55, October–November 2020) writes about the "context, history and evidence which demonstrate plainly that a meaningful number of crop formations cannot be the work of human artists". Frank Joseph (who authored several books about Atlantis) in the article 'The Coronavirus Crop Circles' in *New Dawn* ("#1 magazine for people who think for themselves", 183, pp51–53, November–December 2020) writes "the figures could only have been created by a non-human intelligence outside the Earth." This was particularly about the recent crop circles in Wiltshire in the UK, depicting the COVID-19 coronavirus faithfully replicated from electron-microscopic imagery. This is a bit odd, as the existence of viruses is actively disputed in these camps – for instance, in the same magazine (*Nexus*, pp39–45). Dr. Patrick Quanten in 'The Story of Infectious Diseases' writes that bacteria and viruses are not causing diseases.
33. As discussed by Merlin Sheldrake in the book *Entangled Life – how fungi make our worlds, change our minds, and shape our futures* [2020], and Peter Godfrey-Smith in the book *Other Minds – the octopus and the evolution of intelligent life* [2016].
34. As for instance described by Wally Caruana in *Aboriginal Art* [2003, p14].
35. *Ecology, an Australian perspective* [Attiwill and Wilson (eds.) 2003, pp200–207].
36. [Standley et al. 2009], [Steffensen 2020].
37. A short documentary about the 2012 Cape York Indigenous Fire Workshop can be found on the YouTube channel of CapeYorkNRM (Natural Resource Management), and other projects of The Living Knowledge Place including an Indigenous Perspectives talk by Victor Steffensen at the UTS School of Design on Friday 25 August 2017.
38. [Steffensen 2020]. The Australian Broadcasting Company (ABC) presented a documentary *Fighting Fire with Fire* with Victor Steffensen in the series Australian Story on 13 April 2020, and can be seen on ABC iView and on YouTube.
39. Thanks to Lachlan Falato (2nd year BTI student at UTS in April 2019) for pointing this out to me. The practice is described in an article by Mark Bonta and Bob Gosford et al., 'Intentional Fire-Spreading by "Firehawk" Raptors in Northern Australia', in *Journal of Ethnobiology* [2017]. They provide an overview of eyewitness accounts and relate to traditional knowledge (including indicating that in this context, some knowledge is sacred and that care has to be taken not to expose indiscriminately). Indeed Indigenous academic Tyson Yunkaporta mentions the practice of the 'firebirds' in his book *Sand*

Talk when discussing traditional knowledge systems, of the sparrowhawk totem (in the kinship systems) as a species that "carries burning sticks to spread the fire in the cold season" [2019, p209], and as metaphor [2019, p259].

40. [Pascoe 2018, p166].

41. There is a video on my Vimeo channel which shows a walkthrough of the space created, see www.vimeo.com/bertbon/videos.

42. Wade Davis, *The Wayfinders – why ancient wisdom matters in the modern world* [2009, pp35–78]. I am grateful to Sasha Abrams, MDes student in 2012 who gave me a copy of the book. *The Wayfinders* is based on lectures that Davis delivered in 2009, and was broadcast on radio in several countries by the Australian Broadcast Company (ABC) in 2010. There are also several documentaries by Davis about this topic, for instance *Light at the Edge of the World – The Wayfinders* on the National Geographic Channel, and *Wayfinders – a Pacific odyssey* on PBS.

See also Ulrich Neisser's discussion of the importance of this type of knowledge as 'cognitive maps' in *Cognition and Reality* [1976, pp119–122].

43. *Collapse – how societies choose to fail or survive* [Diamond 2005, pp79–119], the quote is on p82 – I used the 2011 edition.

44. Environmental biologist and historian Jan Boersema, professor at Leiden University, discovered the true situation by reading the original logbook of the 'discoverer' of Easter Island on Easter Sunday in 1722 (hence the name they gave the island, 'Paaseiland' in Dutch, Rapa Nui in the local language) by Jacob Roggeveen, which showed a very different picture than the mayhem as presented in other accounts [2020]. Bregman writes a chapter based on the findings of Boersema, and in an interview Bregman conducted with Boersema the latter exclaimed "I couldn't believe my eyes" (when he first read the logbook), "there was nothing at all about a society in decline" [Bregman 2020, p122], [Bregman 2019, p159] (Bregman in an earlier book followed the 'collapse' theory [2013, pp61–65], as he admits in *Humankind* [2020, p121] / *De Meeste Mensen Deugen* [2019, p159]). From the original logbook of Roggeveen's expedition, Boersema concluded that the inhabitants weren't starving at all, in fact they offered food to the visitors.

Boersema's book *Beelden van Paaseiland – over de duurzaamheid en veerkracht van een cultuur* originally was published in 2011, and after the translation in English as *The Survival of Easter Island* in 2015, a revised edition was published in 2020 (this is the version I read and refer to).

The main conclusion is that there never was a 'collapse' of the society due to overexploitation of the natural resources on Easter Island – but there are plenty of other examples which support Diamond's *collapse* theory, in his several books [Diamond 1997, 2005, 2012].

45. [Davis 2009], [Boersema 2020], [Diamond 2005, p86].

46. [Davis 2009, p47].

47. [Boersema 2020], [Diamond 2005, pp79–119].

48. [Mumford 1961, p13].

49. See Mark Wigley's book *Constant's New Babylon – the hyper-architecture of desire* [1998].

50. *The Empire of Things – how we became a world of consumers, from the fifteenth century to the twenty-first* [Trentmann 2016, p655].

Chapter 5

Industrialising

Science, enlightenment, reformation, and renaissance

Accumulating from the tool period and the settlement period, as discussed in Chapter 3 *Creating* and Chapter 4 *Settling*, respectively, the next historical phase of intense change relevant for understanding the interaction between people and technology is industrialisation, which started around 500 years ago at the end of mediaeval times. And just as in the tool period, where we saw an intricate and complex interrelation between biological evolution, tool making, language, and cultural developments over several millions of years (particularly the last few hundred thousand years), and in the settlement period where we saw the mutually influential developments of agriculture, settlement in villages and cities, specialisation, and written language over several thousands of years, in the industrialisation period there is a similar complex interrelation of changes and developments, over the course of several centuries. And although this development was much faster than previous 'revolutions', it was still a gradual process in many stages and with many interrelated elements, rather than an instant industrial 'revolution'.

Significant transitions took place in religion (the Reformation), emergence of rational scientific methods (reason and investigation), development of engineering (new technologies enabling new uses), new insights in the visual arts and architecture (the Renaissance), and the human-centric world view (Enlightenment and humanism), all of which contributed to industrialisation. In this industrial period there was a shift from manual production to machine-based production, and this fundamentally changed the relationship between people and technology.

The notion that the classical civilisations, particularly the Roman Empire, collapsed due to forces of the northern tribes (Goths and other 'vandals') is although widespread not quite the reality. Perhaps their demise was as much a result of the imploding weight of their own decadence and endless warfare, as much as external factors. There was still substantial development in the Middle Ages (500–1500 AD) after the end of the classical civilisations, as several historians have pointed out, for instance, Lewis Mumford in *City in History* [1961], and to a certain extent also David Landes in his book about the Industrial Revolution *The Unbound Prometheus* [1969], as well as

DOI: 10.1201/9781315373386-5

Ernst Gombrich in *The Story of Art* [1995], who discusses how this notion of 'the dark ages' was in a way a reaction to the feeling of loss of the classical civilisation.[1] In other words, it is more like a nostalgic sentiment than rational reasoning. A similar process can be seen with the Reformation, which resulted in the erasure and eradication of a substantial amount of physical aspects of Roman Catholic culture (paintings, sculptures). There seems to have been an element of propaganda, to suit the ideas of the time, which has lasted until today. Many developments that led to the machine age, science, and industrialisation started in mediaeval times. It is now an increasingly common view that the Middle Ages were not that 'dark' and backwards, that this was partially a view promoted by the Renaissance culture with their wish to return to the classical civilisations. In China and the Middle East, technologically and intellectually advanced cultures thrived throughout mediaeval times. In the Islamic countries, not only did great developments in science and mathematics occur, but they also contributed by retaining classical texts which survived as translations from Greek and Latin to Arabic, and were later translated back again.[2]

As discussed in previous chapters, archaeology, but also history, is by necessity strongly reliant on interpreting (traces of) physical remains of culture, and after the prehistoric times, written sources. Interpretations of finds change over time with developing insights, not all of which is relevant for understanding interaction, so for the context of this book my focus is on the changing relationship between people and technology during this time that relevant developments led to industrialisation, discussed in the following sections.

RENAISSANCE – ART, ARCHITECTURE, AND MUSIC

The Renaissance in the arts was a renewed and fresh start of artistic and creative expression, with a strong reference and restoration of the 'classical'; renaissance indeed meaning 'rebirth' or 'revival' as Ernst Gombrich explains in *The Story of Art*.[3] This included what Gombrich called "the conquest of reality", particularly the invention of perspective painting as devised by the architect Filippo Brunelleschi (1377–1446), reflecting the urge for reason and rationality but at the same time connected to classic ideals. Before this time, the focus of many paintings was religious and the symbolic representation of ideas. In the Renaissance, artists started to depict reality as they saw it, capturing for instance landscapes and objects and actual people – previously people in paintings, such as saints and other important characters, didn't necessarily need to resemble the actual people the images referred to.

In architecture, many new forms and types of buildings started to emerge, but quite prominently here is the return to classical forms and shapes, and integrating these in contemporary architecture.

In music, a number of new developments started, again allowing for more individual and non-religious expression, including polyphony (multiple and simultaneous melody lines) and the invention of new instruments – which in turn led to new musical possibilities being explored. These developments since the early music were ongoing, later leading to baroque music, and classical and romantic musical periods in the 19th century.

SCIENCE – REASON AND KNOWLEDGE

An important development in the industrialisation period was the emergence of the scientific method. The historian Yuval Noah Harari uses the label "scientific revolution" for this period, as this is indeed the key change in his narrative in his book *Sapiens – a brief history of humankind* [2014]. Harari emphasises that during this period, humankind made, as he calls it, "the discovery of ignorance". Although people have always tried to understand the world, around this point in time a significant change in attitude started to emerge and people were willing to admit their ignorance – knowing what one does *not* know is a crucial step towards the potential for increasing one's knowledge, and indeed the basis of all scientific endeavours since. It is in fact, embracing the notion of *doubt*. Admitting ignorance, allowing doubt, is the crucial step towards learning and evolving of insights. Harari describes the scientific method as "the willingness to admit ignorance", and then through "the centrality of observation and mathematics" create the potential of the "acquisition of new power", including insights and new technologies.[4]

As historian Rutger Bregman similarly points out in his 2013 book about the "history of progress", this was the method of thinking as worked out by French philosopher René Descartes in the first half of the 17th century, of *systemic doubt*. This of course doesn't just clash with the dogmatic reasoning of the Church, but also with common sense – where to stop? Inevitably the whole notion of existence of reality is subject to doubt.* In other words, is the fact that someone is doubting, doubtable in itself? As Bregman points out, no, and this is encapsulated in the most famous phrase in philosophy, *I think therefore I am*.[5] Descartes' aphorism is a most effective hack, of the otherwise runaway, logical, circular (or spiralling) reasoning of infinite regress. We do exist, because we ask the question.

The quest for knowledge and insight was also broadened to the wider community. Before science, the knowledge which was captured in books

---

* From Descartes to *The Matrix*, via the Simulacra of Baudrillard, and anticipated by Plato's cave ... this notion of solipsism is discussed elsewhere, in Chapter 9 *Experiencing*.

was often inaccessible to the public, found in the libraries in monasteries particularly, only available for religious scholars.*

New methods were developed which were not based on dogmas, such as those occurring in traditional religions, but based on observation, reflective study, and rational reasoning. The empirical method of science observes, hypothesises, researches through controlled experiments, and analyses the obtained data into solid outcomes. Combined with advances in technology, these explorations were enabled. For instance, the invention of the microscope, by Zacharias Jansen in the Netherlands, led to the discovery of microbes (such as bacteria) in 1676 by Antoni van Leeuwenhoek who perfected lens-making techniques.

Another one of the early milestones which is often cited in this development of scientific reasoning and the approach based on empirical evidence is the astronomical work of Nicolaus Copernicus (born in Polish Prussia in 1473) who mathematically analysed the movement of the planets and other celestial bodies, concluding that the sun (and not the earth, as the religious dogma insisted) was the centre of what we now call the solar system. He published his heliocentric theory in the year of his death in 1543, to avoid religious prosecution. In the next century Galileo Galilei among others developed this theory and calculations further, using a telescope he fabricated himself to make the observations which he shared with others, as A. C. Grayling discusses in *The Age of Genius – the seventeenth century and the birth of the modern mind* [2016]. In spite of actually having supporters in the Catholic regime, including the pope, his many enemies got him in trouble for his heliocentric position and eventually he was convicted by the Inquisition in 1633 – but due to his standing, and advanced age, he was allowed to serve his sentence as house arrest.[6] It is an interesting episode because, as Grayling points out, the dispute wasn't so much about moral or even theoretical insights, but about the power of the Church, which was challenged by the new scientific insights. As Grayling puts it, "Galileo undoubtedly knew that although he had lost a battle, science had won the war."[7]

---

* As the late semiotician Umberto Eco explored in his historical novel *The Name of the Rose* from 1980, set in a northern Italian monastery in 1327. One of the main themes of the book is the monopoly on knowledge stored in books and in this case only in a very limited way accessible to the scholars in the library of the monastery, the monks, and books not agreeing with religious dogmas are locked away. Although presented as a murder mystery and the main character as a detective ('William of Baskerville', a thinly veiled reference to the fictional character of Sherlock Holmes, whose methods of reasoning [Ross 2020, pp44–49] he applies), but more relevant here is the other reference – to the historical figure William of Ockham, applying the early scientific methods. The emphasis in the 1986 movie adaptation directed by Jean-Jacques Annaud, with Sean Connery in the role of William, is on the detective role. But the other aspects are present in the movie too, as well as the example of possible use of optical technology in the form of spectacles, and every time William exposes them in the library reading and writing room (scriptorium), it causes great excitement from his fellow monks.

Science made the previously somewhat vague terms such as mass, force, velocity, light, and time measurable and quantifiable, and manipulatable in mathematical equations and operations, particularly through Isaac Newton's contributions later in the 17th century.[8] (With regards to the measuring of time, I think that it is a particular irony that the calendars changed during Newton's lifetime, potentially shifting the dates of both his birth and his death across the respective dates of the new year.)

REFORMATION – RELIGIOUS PRACTICES

Another important development contributing to the industrialisation period were the changes in religious beliefs and practices. From the 14th century there was increasing awareness and concern about the corruption, both morally and practically, of the Roman Catholic Church – the leading religious influence on everyday life at the time in Europe.*

The Reformation was a response to this, particularly driven by Martin Luther in the early 16th century (then a monk in Wittenburg in the east of Germany), leading to Protestant and other reformed versions of religion. This included the Calvinist tradition founded by theologist Jean Calvin, who was exiled from France, in Geneva later in the 16th century, and in England with the establishment of the Church of England by King Henry VIII after he abolished the Catholic Church in 1533.

All this was a long process over several hundred years, as for instance Lewis Mumford in *Technics and Civilization* writes, including the Counter-Reformation;[9] the burning of heretics, witch-hunting, and other superstitious practices from the 15th to the 17th century. The role of religion in society had a strong influence on industrialisation, through its meticulous organisation; the Reformation supported the later industrial changes through its work ethic and culture. As Lewis Mumford famously

---

* The tension between the pomp and glamour of the Roman Catholic Church and the austerity and piety of the Franciscan monks is one of the key themes in Eco's *The Name of the Rose*, as mentioned, set in 1327. The novel is full of reflections on the tension, which eventually come to a confrontation between the Catholic establishment (and Inquisition) and the Franciscans, and this is particularly well displayed in the movie adaptation from 1986. Another example is the practice of selling 'indulgences', allowing people to pay for forgiveness of their sins as a way to make money for the Church (a practice that clearly goes against the foundations of the religion), and the trade in relics. The appreciation and devotion to relics was still practised in later times, until today. A nice example can be found in Tomasi de Lampedusa's novel *The Leopard* (*Il Gattopardo* in Italian, published in 1958) where there are three spinster sisters who assemble a large collection of (mostly forged) relics in the late 19th century. The later film version by Luchino Visconti from 1963 doesn't show this episode, which is at the end of the book half a century after the main events and a nice reflection (but the film is already three hours long!). The theme of the book is the social changes in the mid-19th century of the revolutions in the process of unifying Italy (Risorgimento), brilliantly summarised in the famous quote *"everything needs to change, so everything can stay the same."*

put it in *Technics and Civilization*, "The clock, not the steam engine, is the key-machine of the modern industrial age."[10] He describes the organisation of canonical hours in life in the monastery, the regular activities, and the adherence to a chronologic structure which is so crucial for regimenting mechanised labour in the factory. (My memories from living in the Netherlands are also of counting the church bells' chimes to know the time, often subconsciously, to contribute to a background awareness of passing and pinpointing time.) David Landes mentions the importance of the clock as well.[11] Accurate timekeeping also played an important role in sea navigation, as mentioned in the section on traditional wayfinding skills at the end of Chapter 4.

Time as a measured and quantifiable entity of course becomes something different from time experienced, as it was before we started using clocks, and is still preserved in some cultures as discussed in Chapter 6 *Communicating*, in the section Language and Time.

HUMANISM AND RATIONALISM

Religion in different forms over history played an important role in people's lives for giving meaning, morals, and structure. As Yuval Noah Harari points out in *Sapiens*, prehistoric cultures adhered to an *animistic* religion, seeking meaning of higher order in the environment including the living (plants and animals).[12] As part of the settlement period, *theism* became the dominant religion, the belief in gods. The scientific revolution was prominently humanist, placing humankind in the centre of the world – with far-reaching consequences. As Harari writes in *Homo Deus*

> Whereas theism justified traditional agriculture in the name of God, humanism has justified modern industrial farming in the name of Man. Industrial farming sanctifies human needs, whims and wishes, while disregarding everything else. Industrial farming has no real interest in animals, which don't share the sanctity of human nature. And it has no use for gods, because modern science and technology give humans powers that far exceed those of the ancient gods. Science enables modern firms to subjugate cows, pigs and chickens to more extreme conditions than those prevailing in traditional agricultural societies.[13]

This is a fast forward interpretation of many relevant developments over several millennia. The responsibility that comes with this humanist position wasn't (and still isn't) fully accepted. The focus is on the 'right' to rule nature, to dominate and cultivate, but we also need to take up more of the responsibility of looking after the planet – if only for our own sake.

In the Christian religion, with the stories and wisdom captured in the Bible, one of the earliest narratives about the responsibility for nature and knowledge was at the end of the Garden of Eden. It was the 'tree of knowledge' that Adam and Eve ate the apple of, and they were punished for their curiosity. This is in contrast to the later attitude of admitting ignorance and allowing doubt, as the means of learning. This became the main approach of the scientific revolution, as historian Rutger Bregman puts it in *De Geschiedenis van de Vooruitgang* ('the history of progress') in 2013: "for the later progressives, curiosity actually became a virtue".[14] Bregman emphasises that in the Bible, curiosity was not encouraged, and also presents the example of Greek mythology which punishes curiosity – Pandora who opened the 'box' or rather the jar,[15] letting out many evils into the world. But at least Pandora was quick enough to close the jar and keep one spirit in: Hope (Ελπιδα).

To take these metaphors one step further, rather than the 'tree of knowledge', could this have been the tree of *responsibility*? Something Adam and Eve didn't have, as no one taught them about responsibility for the environment. It was Eden, after all. Approaching the tree was a sin, an eternal sin, an accomplished fact without any possible redemption. They, Adam and Eve, and therefore *us* collectively as a species, messed it up and that was it. In this interpretation, the biblical version of the story gave us the power, but not the responsibility. Or even the incentive to face the responsibility. However, from Adam and Eve onwards humanity started to learn, perhaps the analogy of L plates (learner drivers of cars) is appropriate – in this analogy humankind currently is still on their (our) P-plates, we have a provisional licence to drive the development of the earth, and unfortunately overall we are not doing very well with this task – yet. This is further discussed at the end of this chapter, in the section on consequences of industrialisation.

In the next sections, this phase of scientific development, of technology and industrial production which involved changes in manufacturing, transportation (of goods and of people), and communication, will be discussed in more detail.

PRINTING PRESS WITH MOVEABLE TYPE

An important part of the transitions which took place from the middle of the 15th century is the invention of the printing press with movable type. Until then, writing and images were produced by hand (by scribes) or by imprint (for instance wood block printing), which had been laborious or repetitive, respectively. Large editions and variety were mutually exclusive. Johannes Gutenberg's invention, presented in Mainz in Germany around 1450, made it possible to reproduce written ideas with variety and in large volumes. Ironically first demonstrated by printing a Bible, as a rather immutable text it was a common item of the previous printing era based on woodcut stamping techniques and handwriting, Gutenberg's printing press nevertheless was a great breakthrough because it allowed to quickly change the content of the text on the page. His invention of 'movable type' allowed changing the individual letters, which were cast in metal (the type, in a lead/tin alloy) and placed in a holder (typesetting). This form of printing was in use as the main form of printing books, magazines, and newspapers until the introduction of offset printing in the 20th century.

It was also the start of the craft of typography, related to the mechanically based presentation of characters. Typography includes notions such as typeface design (assembled in fonts, as occurring on the computers we use nowadays in digital form), and lay-out including line and word spacing, down to kerning (the distance between characters, very important for legibility and aesthetic considerations). This is part of the transition from handwriting to printing, which is the focus of Elizabeth Eisenstein's book *The Printing Revolution in Early Modern Europe* [1983],[16] but it is also part of the larger transition from oral to literate culture.

Gutenberg's invention wasn't just important as a technology and a design tool; this printing press also had relevance as a cultural tool, and it had far-reaching business implications which Johannes Gutenberg perhaps didn't quite manage to oversee, but his partner Johann Fust did. Effectively, while Gutenberg focussed on developing his brilliant mechanism, they also reinvented the role of the publisher. Unfortunately the business didn't work in Gutenberg's favour due to conflicts between inventor and sponsors (as is often the case), but the system invented lived on through Fust's and others' businesses. There was also some resistance of earlier and established cultures, which even led to Fust being identified with Dr. Faust in an attempt to suggest a 'devilish' nature of the invention, in later (and historically incorrect) sources.[17] Apparently the guild of scribes and copyists in Paris managed to delay the introduction of printing by 20 years.[18]

According to Eisenstein, there was also a suggestion that Fust stole the idea from the Dutch printer Laurens Janszoon Coster, who according to the legend invented the movable type technique earlier. There doesn't seem much support for the legend, however of course it is likely that Gutenberg's important invention, with all its complexity, wasn't the work of just one person at one defined point in time. This is often the case, particularly in science, and known as Stigler's Law of Eponymy: "No scientific discovery is named after its original discoverer." This 'law' was proposed by the statistician Stephen Stigler in 1980, referring to the work of sociologist Robert Merton who had stated that "all scientific discoveries are in principle multiples" in his book *The Sociology of Science* in 1973. The eponymy of inventions is therefore often a compromise, and in some cases the most famous or most appreciated person is chosen for the naming.[19]

As stated, printing had been in use for centuries; it can be seen as a mass producing of images, effectively leading to the first notion of mass media. Artists had been using woodcut printing for centuries, and later developed an engraving technique using copper plates (negative and positive printing, respectively).[20] Gutenberg's crucial improvement was the moveable type, so that unique texts could be easily produced (and reproduced), enabling a mass medium for more individual expression, and customisation.* It played a major role in the dissemination of particularly the Reformation ideas; the Gutenberg print technique made it possible to quickly produce many different pamphlets as well as different versions of the Bible, to support the ideas of the reformists[21] (although ironically one of the first purposes of his printing press was the production of indulgences).

Media theorist Marshall McLuhan, writing in the early 1960s, acknowledged this invention as the starting point for the printed medium to become the first mass medium (as reflected in the title of his second book, *The Gutenberg Galaxy* [1962], and also present throughout *Understanding Media* [1964]). McLuhan frequently makes the reference to the assembly line, attributed to industrial practice of the early 20th century, which was a metaphor of printing with movable type. McLuhan also writes about how mediaeval people were still mostly living in an oral culture, with writing a tool for the elite, and the masses being mostly illiterate (while Eisenstein is mostly concerned with the transition from one form of literacy to another). This didn't immediately change with Gutenberg's invention of course. While it is tempting to pinpoint an invention, and glorify its sole inventor, the reality is often different. Usually there many more people involved in innovating roles, either as collaborators or as competitors (as mentioned above), and in many cases a technological breakthrough only becomes culturally significant after further social and economic developments. Mainstream printed media, and widespread literacy, didn't happen until the 19th century. People mostly lived in an aural world until then, as McLuhan observed for instance in *The Medium is the Massage* [1967, p48] (the title paraphrases his most famous aphorism; this book is the result of the collaboration between the media thinker and graphic artist Quentin Fiore – they created a rare example of successful blending of form and content). McLuhan also explored this in earlier publications from the 1950s,[22] in which he writes about the slow transition from oral culture to literate

---

* The importance of this invention, effectively the first example of industrialisation, even before developments in agriculture and in textile manufacturing, is nicely illustrated in the TV documentary *The Machine That Made Us* from 2008, directed by Patrick McGrady [2008] for the BBC. The documentary is presented by Stephen Fry, and the historical aspects are explored as well as the technical – a team of contemporary craftsmen demonstrate how the technologies were developed, from casting the individual types to the actual printing press, inferring what the machine would have looked like (the earliest known image of this type of press didn't occur until more than 100 years later, in 1568).

culture through the millennia-long phase of 'manuscript culture', which was written speech, meant to be recited (not read silently as we have done increasingly since the beginning of print culture), and often memorised.[23] Only with 'print culture', the written mode of expression became more detached from the auditory, and became a whole different medium. As McLuhan summarises, "Phonetic writing reduced the speed of work intake far below the level of oral delivery. Print raised the speed above the level of oral delivery."[24] McLuhan's prime concern of course was to 'understand media', the established and new visual and multimodal media of his time (1960s) such as the newspaper, magazine, and TV, which he approached through the understanding of the older media, by studying the transitions and differences.

Another example which underscores this notion, of how different oral culture was from our current culture with its visual dominance, is that there is only *one* (one!) image of Johann Sebastian Bach, arguably one of the most famous and important composers ever.[25] One image, while his baroque music composed in the first half of the 18th century is represented in an oeuvre of over a thousand surviving compositions, covering a wide range of works including masses, cantatas, concertos, passions, various chamber music works, and scores for individual instruments such as the keyboard and the famous cello suites. The music says much more about the personality of J. S. Bach than the visual image. Similarly, there is little known about William Shakespeare as a person, who lived more than a century earlier, and the only images of him are dubious attributions. There are several images of Johannes Gutenberg though (but not of his printing press, it is not actually known what it looked like).

It is interesting to note that in earlier times images of a famous (or in fact, any) person didn't necessarily even attempt to accurately represent what the person looked like. It just wasn't considered necessary. Fidelity became an issue much later in history, in painting, and of course irreversibly accelerated with the introduction of photography and film in the 19th century. This is why we must appreciate the radical nature of the complete anonymity of graffiti artist Banksy.[26] Though not all graffiti is art, some of it is perhaps vandalism, I have seen some of the strongest artistic expressions in the last decades in this medium. I consider Bansky one of the most conceptually interesting contemporary artists, and some of his work at the same time has pop appeal (that stencil art of the girl with the red balloon is everywhere), yet the public does not know who he is or what he looks like. While images of his work (mostly his street art) are extremely popular, he is defying the rules of popular culture of the importance of the visual image of the artists (in contrast to J. S. Bach's time). By not surrendering to this rule, there is more focus on his images and his activist messages and his work rather than on his own image or him as a person. Of course there are other (pop) artists who manage to remain at least some level of anonymity, hiding behind make-up (the hard rock band *Kiss* in the 1970s), props such as helmets (French techno duo *Daft Punk*), or a mask (guitar virtuoso *Buckethead*).

INDUSTRIAL REVOLUTION

Just as there was light in the 'dark ages', there was darkness in the Enlightenment. In spite of all the lofty ideals and new insights in science, humanism, and rationalist attitudes, there were many wars and civil unrests in the period leading up to industrialisation. After a century of conflict, mostly between the Protestants and the Catholics particularly in the Thirty Years' War (1618–1648), which Ernst Gombrich describes as "terrible times,"[27] in the second half of the 18th-century mechanisation of manufacturing accelerated with the Industrial Revolution. An important invention was the steam engine, which took over as an increasingly powerful source of kinetic energy. First wood was used as fuel, later coal which created an insatiable demand for this material to be mined.

The historian and economist David Landes in his book *The Unbound Prometheus – technological change and industrial development in Western Europe from 1750 to the present* [1969] identifies three key areas of material change: human skills being replaced by mechanical devices, human and animal strength being replaced by 'inanimate power' (primarily steam), and improvement in processing materials (metallurgy and chemistry).[28] Particularly the first two areas had a major impact on the interaction between people and technology, leading to new social (or rather, anti-social – labour became deskilled and therefore cheaper, and this was often exploited by industrial leaders) structures, and new business models. Improvements in agriculture played a role in this too. For the first time, machines started to play a more central role than humans. Machines took over in textile production, and soon in other areas of manufacture too. Complex mechanical technology had to be designed in a way that it could be operated by people. However, the people often played a subservient role in this relationship, and this is important for understanding the history of interaction. This subservient role of people in relation to the machine was particularly strong in the mining industry. The degradation of the worker was to a level of 'wage slavery', while the work itself was very dangerous.[29]

As the film maker Humphrey Jennings puts it in his book *Pandæmonium – the coming of the machine as seen by contemporary observers*, "In the two hundred years 1660–1860 the means of production were violently and fundamentally altered – altered by the accumulation of capital, the freedom of trade, the invention of machines, the philosophy of materialism, the discoveries of science."[30] The title of the book refers to Milton's epic poem *Paradise Lost* from 1667, Pandæmonium* is the name that Milton made

---

* Note the old British spelling of *Pandæmonium*, I actually considered to use the æ character throughout these chapters, for words like palæology and archæology to restore the role of the seemingly obscure characters, but it looks a bit too odd and besides the red underlining in MS Word is just too annoying in the editing process. Still, I want to defend the use of other characters than just the 26 in the alphabet, like ö, ô, ø, (they're like smileys!), ß (double s in German), Greek alphabet (α, β, etc.) characters. I thought Elon Musk was agreeing, but in 'X Æ A-Xii Musk', apparently Æ is Elven spelling for AI.

up for the capital of hell. The book is often described as a narrative 'telling the Industrial Revolution in real time', and indeed the selection of writings Jennings collected and annotated, presented in chronological order, give a close account of the period of industrialisation from first-hand sources.

The Industrial Revolution as such started in Britain, based on the earlier transformations of cotton manufacture that took place early in the 18th century and which gave rise to the factory system as a new way of production.[31] That it happened in Britain was for a number of reasons: this includes *geology* (the presence of coal which could be mined), *geography* (rivers that were navigable, later canals were constructed), *agriculture* which was increasingly industrialised or machine-based, and all the factors and developments described above of the Enlightenment, scientific and technological advances, and religion. In relation to the latter, the English king Henry VIII had abolished the influential Catholic Church in the 16th century (in 1533) and had taken the first steps to establish the Church of England, a reformation of religious but also material nature, as it led to a redistribution of wealth.[32] (England did not participate in the Thirty Years' War.) The Reformation on the continent of Europe also had had an influence on the balances of powers. Recent research suggests that the abolishment of the Catholic Church in England, the dissolution of monasteries, and the redistribution of assets (land and buildings) to new owners correlates statistically with locations of industrial activity and wealth in later centuries.[33] This dissolution would have had a role in the Agricultural Revolution, as it led to a rise of a gentry class (non-noble landowners) of larger scale farmers. Another important factor in England was the lack of internal tariffs which encouraged free trade. In other countries, such as Germany which was still a fractured nation at the time, but also in France, the industrial developments were delayed due to tariffs between regions which obstructed the trade. In France this lasted until the French Revolution, which started in 1789.

To illustrate the importance of geography and geology, it is interesting to note that after Britain, the second area where the Industrial Revolution took off was Wallonia, the southern and French-speaking part of Belgium (which had separated from the Netherlands in 1830), from the early 19th century, due to the presence of coal and navigable rivers. By contrast, previously the northern and Dutch-speaking part of Belgium, Flanders, was prospering with the mechanised textile industry (particularly the town of Ghent) before the wealth moved south to Wallonia. The Netherlands, meanwhile, missed the boat, as their main source of fuel was peat ('turf'), which was not suitable as a fuel for steam engines, and also lacked iron ore.[34]

Industrialisation was at first a development of scale, but also of scope. There was a change from the 'cottage industry' where skilled labourers, craftspeople, and farmers would work directly with their materials and processes. After farming, the next change was in the textile industries where processes were mechanised, and finally new industries emerged.

The machines were expensive and could not be owned by individual labourers, leading to the development of factories to support the necessary scale of operation. The role of the workers in the factories was very different from that of the independent workers previously. Most of the work was unskilled, which meant there was much more competition for work, something the factory owners exploited by lowering wages. Poor working conditions and low wages led to a lot of unrest, protests, and even attempts to sabotage the machines. The Luddites and other initiatives were mostly directed at the exploiting factory owners, rather than the actual machines (although this is what they approached). But as Ernst Gombrich notes in his chapter 'Men and Machines' in his book *Little History of the World* (originally written in German in the late 1930s), "in England in 1812 there was a death penalty for anyone guilty destroying a machine."[35] This inescapably suggests that machines were seen as worth more than people. William Blake famously wrote in *Milton a Poem* in the early 19th century about the "Dark Satanic Mills".[36] Earlier, an also quite well-known insight comes from the historian William Hutton who recollected memories of his time as a child labourer in the Derby Silk Mill ca. 1730, and reflecting on this Humphrey Jennings points out in *Pandæmonium* that life of the poor people was confined to one building which served as factory, school, workhouse (a place for forced labour), and prison.[37]

Eventually this led to the people uniting in trade unions, to claim some power which they couldn't have as individuals. This socialist trend as a response to capitalism was the basis of later communist systems.

I am deliberately emphasising the negative aspects of industrialisation. Although much of the Western world as we know it was created in this period, and we owe our (material) wealth to this development, we are now at a point in time where we start to become really aware of the negative effects of the system, where it starts to actually work against us through damage to the environment. A progress trap, as it is often called.[38] The flaws of the capitalist system, consumerism, and the fallacy of eternal growth (an obvious contradiction if one thinks about it), the main elements of the crises we are in now, all started in the period of industrialisation.[39]

TECHNOLOGICAL DEVELOPMENTS DURING INDUSTRIALISATION

In this section the developments are explored further, to show how the relationship between people and technology and the interaction with our environment changed over time.

As mentioned above, another way to look at it is to see the whole system of technology, bureaucratic organisation, and power structure, as a machine. In a sense this is what Neil Postman does with his terms 'technocracy' and 'technopoly' [1992], and earlier Lewis Mumford with what he calls the 'mega-machine' (using the building of the pyramids as an example, it is the whole complex of technology, material, organisational structure, and people – mostly forced labour) in *The Myth of the Machine* [1967]. And of course, this is what the Church was defending in the period of the Enlightenment, not so much the truth, but rather the *power*, in the form of their own bureaucratic structure which ruled society. The famous case of Galileo *vs.* the Roman Catholic Church in 1633 illustrates this, it was never about the quest for truth – it was inevitable to accept the heliocentric view-point – but it was about power and influence, about authority.[40]

Different technological developments played a role in these shifts in society, in the social and organisational structures. In the next sections, the development of **active mechanical technologies**, and later **electrical technologies**, are further discussed.

From passive to active mechanical technology

During early industrialisation, steam engines took over most of the work previously achieved by water pressure or wind, and animal power. Later the combustion engine (gas, diesel, or petrol) increased versatility and power, applied in the machines.

The internal combustion engine uses a piston inside a cylinder, driven by the rapid expansion of an exploding amount of gas in a confined space (by an electric spark, and/or by sufficient compression, as in the diesel engine). The earliest example of this principle is actually a gun, but using gunpowder proved unsuitable as a technology to create and engine.[41] From the mid-19th century a mixture of gas and oxygen was used as fuel. Later, liquid fuels were used, which were easy to store and transport.

In the classification of technologies introduced in Chapter 2, these are examples of **active mechanical technology**, in combination with **chemical technology**. (I am focussing on combinations here, although chemical technologies have had a massive industrial impact and the interaction between chemicals and people can be profound, in medicine and other applications). The earlier active mechanical technologies were either using kinetic energy (water pressure, wind) and were therefore entirely mechanical, or using animal or human power. It should be noted that water- and wind-based kinetic energy is usually entirely free, which has an influence on the way the energy is used, in contrast to the engines invented later. For centuries, windmills were used in the Netherlands to pump water out of parts of lakes, by putting in dykes to separate certain areas to be turned into dry land ('polders') to be used for agriculture or building houses or industry. The national airport is on reclaimed land, reflected in the name 'Schiphol', 'ship hole', because of the ship wrecks that emerged after pumping out the water.

In contrast to wind and water, animal- and human-based kinetic energy is costly. In this latter form, there is the particular case where the human power is delivered by the same person who is using the structure, such as in a pedal driven sewing machine, or a bicycle (still popular!). In this case, the delivery of power (such as pedalling) becomes part of the interface between person and technology in a direct control mechanism. In all the other cases, there are differences (however subtle) in the way the machine is operated depending on whether it is powered by animal, kinetic, or chemical (and later electromagnetic – electric bike!) power.

A special situation is the case where the human kinetic energy is stored in the mechanical structure, for instance by winding up a spring (as in a watch or clock). This detaches the energy delivery from the act of operating the machine. As long as the stored kinetic energy lasts, it gives the same impression as working with an engine – the interaction is similar. Not much energy can be stored this way, however, so this technique is mostly used for low energy processes such as the ones enabling communication – particularly clocks.

From electrical to electronic technology

The influence on industrial processes through the development of the internal combustion engine and further developments of electrical technology that took place towards the end of the 19th century are often described as the Second Industrial Revolution, for instance by David Landes, and Rutger Bregman who points out that these developments were led in particular by Germany and the US.[42] However it seems mainly an incremental change rather than a revolution, all the elements were already in place but the effects magnified due to technological advances. In the section below in this chapter, Phases of Industrialisation – the Unindustrial Revolution, this will be discussed, following from the framework of technological functions.

During the period of industrialisation not only do we see a change in production and manufacture, but due to the transition from purely mechanical technologies to the inclusion of chemical, magnetic, optical, and electrical technologies, we see an increased presence of the functions of transportation, transmission, and communication, discussed in sections below. Firstly, we need to look into the developments of **electrical technologies**, including the later more sophisticated **electronic technology** (which is about changing and modifying the electrical signals, first in **analogue electronics**, later **digital**).

Electricity occurs in nature, such as lightning and static electricity, which is generated through friction. As such, electricity was known to humankind, but only since the late 18th and early 19th centuries was it studied scientifically, and its behaviour and possibilities for applications understood.[43] Rather than an obscure and possibly magical property, it was realised by the early pioneers (most of their names live on in eponymous units and laws) that if there was an electric charge, it was a potential to do work, nowadays known as **voltage**, potentially resulting in a **current** which is a flow of electrically charged particles. These charged particles are usually electrons, but can also be ions, which are atoms that are electrically charged (positively or negatively). Electrical processes in organic matter including human beings are often based on ion movement, rather than on electrons.

Voltage, with the unit volt, or V (named after Allesandro Volta, 1745–1827), is defined as the potential difference (between two points in a circuit), or the electromotive force. It is useful to think of it in these terms; in other languages this is reflected in the word for voltage, for instance in Dutch voltage is called 'spanning' which means tension. A connection with a certain electrical **resistance** (in ohm, or Ω, or R) will result in the flow of charge: the current, with the unit ampere, or A (named after André-Marie Ampère, 1775–1836). Resistance is the opposite of **conductance** (G, which is 1/R). The relationship between these three basic terms is described in the famous Ohm's Law (after Georg Ohm, 1789–1854), which defines the electrical resistance (R) as the ratio of the voltage (V) and current (I): the formula is R = V/I, which means that if two of the values are known the third can be calculated. A further useful relationship is in the definition of electrical **power** (P, in watts, W, after James Watt, 1736–1819) as the product of voltage and current (P=V x I), as this relates to the measure of how much energy is consumed over time (W/h).[44] This episode shows the strength of the scientific reasoning, by discovering the laws that describe and predicting the electrical phenomena, and then applying them in technologies developed in engineering.

To understand electricity, the analogy of flow of water is often used: as a result of potential energy (water stored in a bucket, raised – the voltage), a flow of water (the current) can occur, with a speed determined by the restriction of the water pipe (the resistance). This of course only holds with constant flow (DC, direct current in electrical engineering terms); in the case of changing currents (AC, alternating current such as the mains supply, and audio and sensor signals) it gets a bit more complicated. The 'alternating' is usually in the form of a sine wave (in the case of electrical power systems, due to the way it is generated) or more complex waveforms, with a **frequency** which describes how often the signal changes in a second. Frequency is expressed in the unit hertz or Hz, the number of cycles per second (named after Heinrich Rudolf Hertz, 1857–1894 – not the car rental company).

Electrical batteries (or voltaic piles) became available for these experiments in the early 18th century, combining **chemical** and **electrical** technologies. Another important combination of technologies is for example the basic principle of the relationship with **magnetism** and **electricity**, which is a mutual influence. So, if an electric current is flowing through a conductor such as a wire, it produces a magnetic field around it (as first observed by Hans Christian Ørsted early in the 19th century), proportional with the changes in the electrical signal. And vice versa, as observed by Michael Faraday a few years later, the converse effect also occurs, a magnetic field can 'induce' an electric current in a conductor. These reciprocal relationships were captured in the formulas that James Clerk Maxwell formulated around the middle of the 19th century.[45] Relating this magnetic change to a permanent magnet results in a kinetic (mechanical) action, and conversely, a moving part in an electromagnetic set-up (a moving electrical conductor in relation to a static permanent magnet) will result in an electrical current (due to the displacement of electrical charge), a flow of electrons, an energy potential. In order to amplify these effects to workable proportions, the wire (electrical conductor) has to be long, and is usually rolled up in a *coil*. This principle is applied in electric motors (converting electricity in kinetic mechanical energy) as well as in generators (converting kinetic energy in electricity), and is still around today. Another example of the application of the electromagnetic principle is in devices relating to the kinetic energy in soundwaves, in (dynamic) microphones and speakers.

Energy converters are called **transducers,** and they are important for understanding interaction because they play a crucial role in connecting the outside world (including human beings) to technology. There are two types of transducers: **sensors** which transduce energy into a machine, and **actuators** which facilitate machine outputs. In the earlier example, a microphone is a sensor, which converts the kinetic energy of moving air particles of sound waves into analogue electrical signals, and the speaker is an actuator that converts electrical signals into analogue kinetic energy (sound waves).

Early examples of electrical technologies are the telegraph, the first telephone, and the light bulb. Later, electronic technology was developed. **Electronic technology** is electrical, but it has the ability to change and modify electrical signals, with the invention of the vacuum-tube or valve in the early 20th century. An example was the *diode*, which conducts electricity only in one direction, which enabled the controlling of signals.

Analogue and digital electronic technology

Electronic technology can be divided into **analogue electronic technology,** where the electrical signals follow the energy transduced (analogously), and later **digital electronic technology,** and finally the digital **computer.**

Digital means 'through numbers', not the same as **binary,** which is the restricted situation where only two numbers are used (0 and 1), as electronic switches (on/off) are the basis of many digital circuits (for storing information in binary code, and for processing). Binary numbers can represent any number, if there are enough digits. Our decimal number system is based on the number of fingers we use to count on – growing up with that as a child makes it seem intuitive, but it is effectively a convention. This decimal (base-10) system is a learned thing, mathematically speaking the number 10 is an arbitrary number. We could easily use another system; earlier cultures counted the individual digits of one hand (hence the term digital, *dig*). Using the thumb to point at the individual digits of the four fingers, this technique enabled counting past 10, it was very common to count to

12 or 16. Examples of the use of the base-12 counting system (duodecimal or dozenal) are in terms like dozen (12, and 144 – a gross which is 12^2), the twelve notes in an octave (equal intervals, the pitches between the doubling of frequency), 12 months in a year, 24 hours in a day, 60 seconds in a minute (there have been attempts to introduce decimal time, which seems very confusing at first, showing how much it is about what one is used to!), and in Britain until 1971, 12 pennies or pence in a shilling, and 20 shillings in a pound sterling. In many languages there are unique names up until 12, and some even up until 16 (French), which indicates that the decimal system has not always been the norm. These different ways of counting on the fingers or digits of the hand were enabled by the evolutionary development of the opposable thumb (in early humans, and some apes have this ability too), and early modern humans developed further dexterity which enabled them to touch all available digits of the fingers by reaching with the thumb. This principle of co-evolution in human development is discussed in Chapter 3 in the section Composite Tools, on the interrelation between bipedalism (walking upright, which freed the hands from their task of (knuckle) walking and climbing) and the emergence of sophisticated tool use. With this example, the interrelation in evolutionary developments is further extended with the influence it had on numeracy and language.

Computers often work with the hexadecimal system (because it is more efficient when based on binary coding), which is base-16 (1-9, A-F). So, where we are used to counting to 10 and every time we reach past 9 we add 1 to the next position to the left (increasing the 'place value'), in hexadecimal a person (or usually the computer) counts on after 9, using the letters A (10 in decimal) to F (15 in decimal), until you get past 15 (F) when you add 1 to the next position. Binary, the base-2 system, shifts much quicker, as soon as you reach 1, the next step adds 1 to the next column to the left (each with a place value of the next power of 2). Starting from 0, then 1, then 10, then 11, then 100, etc. (On a bushwalk once I saw someone with a T-shirt that read "there are only 10 types of people in the world: those who understand binary, and those who don't.") (Another one is "binary, it's as easy as 01, 10, 11.").

Again, if the numbers are large enough, a high precision equal to 'analogue' electronic or even better can be achieved. Eight bits are a byte (0-255), and sixteen bits (two bytes) are enough to translate an (analogue electrical) audio signal into data if sampled at a high enough frequency (the sample rate is 44.1 kHz in the case of the CD, chosen to be more than double the highest frequency the human ear can detect, 20 kHz (but in practice it is much lower, as discussed in the section on auditory perception in Chapter 8 *Feeling*), the Nyquist frequency is 22050 Hz); most computers now work with 64-bit numbers. Today's streaming of digital audio content is often of slightly less precise coding, through the use of data compression (the process of taking out redundant coding information (lossless conversion) which results in less data to be streamed, and in other cases data is lost which can be audible, leading to a lesser quality of the audio signal). These techniques give digital a bad name, but in fact the technology is deliberately limited for cost saving reasons.

Early **computers** were mainly used for number based operations, 'computing', which was until then carried out by humans ('computer' was a job title[46]). But the truly distinguishing feature of the computer is that it can be programmed, through software. And because of this ability to change its function instantaneously, I consider it a separate category of technology. The first computers were actually mechanical (such as the loom, and Charles Babbage's programmable 'difference engine' invented around 1830, and his later 'analytical engine'), later using analogue electronics and eventually using digital electronics.

With the invention of semiconductor technology, the 'solid-state' or semiconductor devices in the middle of the 20th century, such as the diode and particularly the transistor (invented in 1948[47] and becoming mainstream in the 1960s), electronic technology became more miniaturised and embedded in many other technologies. While the vacuum-tube had already enabled new functions and even new media such as an improved version of the radio, telephony, amplification, and the television (and new musical instruments, which were developed to make electronic music[48]), due to the size of the tubes and associated power circuits the technology was rather bulky, had a limited life-span, and used a lot of electrical energy. Most of the energy used was turned into heat, a by-product of the vacuum-tube, indeed a kind of light bulb. And where vacuum-tube-based computers would easily fill up a whole room, transistor-based computers fitted in a cupboard-size space, and were more reliable too because a transistor lasts much longer than a vacuum-tube. The transition from tubes to solid-state took several decades though, and vacuum-tube technology can still be found in some niche areas. Examples are tube-based high-end audio amplifiers (because they amplify the signals in a different way, the sound is often described as 'warmer', possibly a misattribution based on a sentimental connection with the high temperatures generated by this technology) and in guitar amplifiers – in this case, because when using the signals in overdrive the resulting distortion with a tube is different from the transistor, and because distortion adds higher harmonics to the signal the pattern actually is different, and often preferred.

Semiconductor technology, first using germanium, later mainly silicon, allowed for further miniaturisation by packing multiple transistors to the same substrate, the *integrated circuit* or IC (or chip) invented in the late 1950s and becoming widespread in the 1970s. The density of components on the silicon chip has been rising exponentially, leading to exponential increases in processing power and storage capacity. This is represented in Moore's Law (after the founder of the Intel chip company, Gordon Moore, who first predicted this in 1965), stating that the number of transistors in an integrated circuit doubles every two years. As anyone who has been using computers for some decades will have noticed, this doesn't mean computers are getting exponentially faster and more powerful, due to software getting more complex, at least at an equal pace it seems. In fact, having used computers for over thirty years, my experience is that in some aspects they seem to have gotten slower and more sluggish in the interaction in the last decades. I keep a few old computers running, to check that this isn't subjective; it is known that with ageing, time perception changes, and it isn't *me* that is thinking faster as I age – or just becoming more impatient.

This is further reflected on in Chapter 12 on design perspectives and applications.

TECHNOLOGICAL FUNCTIONS

Near the end of the 19th century and at the beginning of the 20th century, many of the functions of technology, as introduced at the end of Chapter 2, were in full use so this is a good point to reflect on them in the historical context.*

The main functions, related to power and information, are **transport** or distribution and delivery, generation or **conversion**, and **storage**.

Power and energy sources

As mentioned, it is useful to make a distinction between technologies used as energy sources, to power the manufacturing and other processes, and technologies for information and communication. In the next sections these two main functions are covered, first for power, and then for information.

Transmission and distribution of power

The early factories generally relied on a source of kinetic energy, first wind or water pressure, but as discussed above the effectivity was greatly increased with the introduction of the steam engine. The kinetic energy was distributed throughout the factory or workshop, using axles and belts.

Later, with the internal combustion engine which eventually became smaller, more versatile, and wider distributed, it became more common to have multiple engines in a factory or workshop, one for each different

---

* In fact, this is what the cyberpunk writers William Gibson and Bruce Sterling illustrated with their novel *The Difference Engine* from 1991. This science fiction story, set in the first half of the 19th century (!), envisions what *could* have happened with all the inventions which were already existing though not fully developed, such as communication by fax machines (!), and the designs of Charles Babbage of the first programmable (mechanical) computer, the machine of the title (and also extrapolating the role, based on the historical fact, of Ada Lovelace's friendship with Babbage and her being the first computer programmer).

purpose. This principle was further developed with the electrical engines. The notion of the central source of kinetic energy is worth reflecting on, as it is very different nowadays where every tool in the workshop (or in the kitchen) has its own motor, a decentralised situation which has many advantages.

Electrification started the now common practice of an electricity 'factory', a power plant or power station generating the electrical power (converting from mechanical, kinetic energy) which then gets distributed over longer distances. The choice for AC (alternating current) is related to this, because higher voltages (several tens of kilovolts) travel better over long distances, and with AC it is easier to change the voltage (through transformers, as an electromagnetic device using the principle of conversion through electric coils and interrelated magnetic fields) so that voltages can be used to distribute electrical energy over long distances, and relatively low voltages locally (120 V in the US, 240 V in Europe). A popular application of this system was the electric light, in streets and domestically, replacing the earlier mains gas-based distribution of energy for light sources.

Another form of distribution of energy can be found in *hydraulic systems* which use water pressure (or other liquids, particularly oil), and still in use. Earliest examples can be found in the factories and workshops that were driven by water power (from rivers or streams), and the aqueducts from the Roman era.

Pneumatic systems use (compressed) air to transmit energy. This technology has certain advantages and is still in use in factories and workshops (many pneumatic power tools can be found there). Usually there is only one centralised source of kinetic energy (the generator that compresses the air) which then gets distributed throughout the workshop, just like the situation in a workshop driven by a steam engine, or earlier, water or wind power (water and wind mills), unlike the situation with electric energy (many battery powered tools are in use all over the shop). Pneumatic and hydraulic systems are sometimes used in situations where electricity would be dangerous (due to the risk of sparks, in flammable or explosive environments).

Storage of energy

An early example of the storage of potential kinetic energy is the water tower, which is a source of continuous pressure to be applied in many ways. On a larger scale, this principle is also applied in reservoirs and dams, containing water at a certain altitude, the (mostly constant) pressure of which can then be used particularly to drive turbines for electricity generation (hydropower).

Other techniques for storing kinetic energy are the spring, and the flywheel (also to even out fluctuations in kinetic energy delivery), and compressed air in cylinders.

Electrical energy can be stored through chemical processes in a battery.

Wood can be regarded as one of the earliest examples of stored energy, which can be released through conversion into heat by burning. The ancient practice of carbonisation involves partial burning and removing the water from the wood, through limiting the oxygen supply to the burning process. This creates charcoal, which burns better than wood, it is lightweight and can be used at a later stage when needed.

Conversion and translation of energy

An important aspect of technology is the *conversion* of one form of energy into another. We tend to think of this process as 'generating energy', but what we mean by that is generating *useful* energy – energy is never generated (or lost), only converted. This conversion is generally from a less useful form to a usable form of energy, for example, heat into electricity. When we 'consume' energy, we turn a useful form of energy (such as electricity) into a less valuable form of energy (such as heat – in the long run); in the short term it yields an advantage. In a closed system, energy never disappears, it only gets converted. This is the first law of thermodynamics, where the second law of thermodynamics states that with each conversion the energy becomes less useful, less valuable, less available, and not reversible – this is called 'entropy' (by Rudolf Clausius in 1865).[49] The often occurring irreversibility is something to watch out for, as it means that with each conversion the energy becomes less useful.* This means that the entropy increases (an increasingly inescapably cleart hat this is contributing to climate change).

During industrialisation, burning fuel was an increasingly common way of releasing energy stored in matter, chemically into kinetic energy, first using steam and later with the internal combustion machine. Translating of movement is a common aspect, reciprocating motion (linear, up-down) of the engine into rotary motion, which is needed to drive axles, wheels, and cogs for gears (further translating of torque and rotational velocity). The earliest example of a translation of kinetic energy would be the *lever*, a rigid rod on a pivot, which can turn a large movement with little force into a small movement with a lot of force, and vice versa. Presumably this is a technique used for building early polypod structures such as dolmen and pyramids.

The invention of the turbine was a great improvement in efficiency in the mechanism of converting kinetic energy, still in use for electrical power generation.

---

* Which makes time travel highly implausible, not only because of the general paradoxes (explored in movies from *The Terminator* and *Back to the Future* from the mid-1980s, to *Tenet* in 2020), but also because of the tensions with these laws of thermodynamics. Hence the use of the term 'inverted entropy' in Christopher Nolan's *Tenet*.

Information and communication

In addition to the existing ways of using technology as a power source, and as a means of producing goods, technology since the industrialisation period is increasingly used as a means of communication of information through the transmission of signals (and later digital data) and storage of signals and data (recording).

This is sometimes called the 'Third Industrial Revolution', but that is a bit confusing as it is a shift of functions, leading to a different paradigm. This development is about media, rather than manufacturing, and particularly prominent since the mid-20th century. This is discussed in more detail in the next chapter, Chapter 6, which looks at communication and information, but in the sections below some of the technical and historical aspects are discussed.

Transmission of information

Although (long) distance communication has long existed by using mechanical technology and other means, such as semaphore (expression at a distance using flags[50]), smoke signals, and sounds (see James Gleick's fascinating account on the African talking drum as a long-distance communication medium, in *The Information*[51]). Electricity turned out to be a great breakthrough in this area. Electrical signals could carry information over a long distance, and with the development of suitable transducers it was possible to turn sounds into electricity (microphone) and vice versa (speaker). The first electric telegraphs were able to relay kinetic messages over a distance, further improved with the development of Morse code[52] and later sound-based communication. The telephone facilitated the electrical transmission of the voice.

The purpose of purely mechanical technology and the mechanical and chemical technology combination in the machines as described earlier was mostly about providing the driving force of the needs of industry, for manufacturing. Electrical technology, although playing a modest role in the same realm (electromagnetic technology in motors), was particularly suitable for applications in communication as the mediation of human thoughts and intentions. By converting into electricity (transducing) events in the real world, such as an utterance expressed with the human voice or any type of the previously ephemeral events in the sonic realm, it became possible to transmit this utterance, this event, this information, through electric wires from one location to the other, over increasingly long distances (and eventually without: *wirelessly*, through radio waves, early radio was called 'the wireless'). And subsequently with the invention of recording media it became possible for sonic events to be stored permanently (first mechanically, capturing the sound waves into matter like wax-rolls and discs, later electromagnetically – though the actual technological developments in this progression were a lot more diverse).

This is an important shift in industrialisation, from manufacturing to communication.

This potential was probably more important than the industrial powers facilitated by the electric motor, and the generator, and the role they played in increasing manufacturing productivity. The extended communicative power of the recording and transmission of human expression, enabled by these technological developments, had far-reaching consequences. This was already identified in the 1960s particularly by Marshall McLuhan's notion of the Global Village,[53] all the way up to the Internet and all the new (social) media that emerged with it.

Electromagnetic transducers – the microphone and the speaker – when electrically connected facilitate the transmission of human expression over a long distance, in real time, 'broadcasting' to many receivers. This enabled individual orators to talk to many listeners, through amplification with Public Address systems in the same area, or over longer distances using wires and later through radio waves, detaching the orator from the listener.

With the early telephone system, connections between people with the intention to communicate were made first through a human-operated switchboard (one had to ask who one wanted to speak to) and later by inputting in a number unique to each individual recipient. It may be good to reflect on this, as in our current age of smartphones one rarely keys in a number, let alone memorise one, but not long ago people actually memorised several, up to ten-digit, phone numbers of friends and family. (Being part of that generation, I have this story: my mother recently got rid of her landline – we nowadays always use internet-based media to communicate – I actually regretted not ever again needing the knowledge of the full fifteen digits stored in my memory, the last seven digits being the local number and the most significant.) One could also look up names in a phone book which listed every person with their number (unless they opted out being listed).

Before this interface though, early phone technology was entirely speech based (interaction with the operator to request the connection with a person took place in the same modality (speech) as the actual communication). The later phone technology added the interface which allowed the person to select the number of the recipient – first with a rotary dial, then with numerical key buttons (still emulated in the smartphone on-screen interface, but you can actually use a rotary dial emulator app).

Recording – storage of information

After the media of writing (as discussed, capturing information first in clay, stone, papyrus, parchment, and eventually paper), and recording information mechanically, the new electric-technology-based media soon led to ways of storing information in electromagnetic form.

Electrical technologies had a strong influence not just on the industrialisation of labour, but even more in the area of communication. Since the invention of writing, it became easier to record ideas, insights, and knowledge. Until then, in oral culture, knowledge was captured through memorising, including storytelling, dialogue, and regular repeating. In oral culture the need to memorise has had a strong influence on the way narratives are constructed and presented, and supported by physical objects and even architectural spaces (the mnemonic technique of loci), which Lynne Kelly in *The Memory Code* [2016] calls 'memory spaces' – illustrated at a range of scales. With drawing and writing it became possible to capture certain aspects of events and ideas – but it was always a representation, and more static than oral representation. Oral cultures can be more dynamic than literal cultures, under constant revision as Kelly points out in *The Memory Code*,[54] and less prone to bias towards certain aspects of the information captured. Australian Indigenous scholar Tyson Yunkaporta makes this point too; in his book *Sand Talk* he compares the two cultures from his personal experience, oral as dynamic and literal as static.[55] This is further discussed in the next chapter, *Communicating*.

This process of capturing through representation (writing) was also applied in music, which previously was ephemeral, the sounds were gone as soon as they were produced (as physical entities, of course the notes, the music, the meaning would potentially linger and resonate in the minds of the listeners for much longer). But by writing down the notes in a score (sheet music) it was possible to reach new levels of complexity. Like writing in language though, it was a *representation*, not the actual music and by necessity this process would leave out information. Even the intermediate stage which included instruments such as the player piano, the information was coded as holes in paper which when fed through the machine would make the keys 'play' the notes according to the score, resulting in the sounds representing the score. When it became possible to actually capture the sound as it was produced (on recording media such as discs and tapes), either as speech or as a musical expression, this was a great breakthrough. First this was done in the 19th century with purely mechanical means, capturing the air waves (variations in air pressure) with a mechanical transducer in a trace on a wax roll, which then could be replayed by a needle following the track and mechanically amplified with a horn to audible air waves. Later this process was improved with electromagnetic means, with increasing fidelity (but never with the full richness of the original acoustic events, though some hardcore audiophiles might disagree). And again, as with language, the dynamic nature of improvising, or any form of live music, is very different to the static nature of the recording. This is why a new way of making music was developed, from the electronic avant-garde in the 1950s to recent developments in techno and other 'electronica' music – clearly in this idiom the recording is the actual work, which may be followed by dynamic live performances, further variations and developments, including remixes.

TECHNOLOGY APPLICATIONS AND RESPONSES

Industrialisation encompassed a further range of changes in society. In the next sections a number of key developments are presented: transportation of goods and people, extensions of industrial manufacturing applications, and the various responses to them.

Transport

In the period of industrialisation a major change in transportation occurred, of trains, ships, and cars using the technologies described above. Steam trains started to appear around 1800, while earlier horse-drawn carriages on rails were common. Ships, which were until then propelled by human power (rowing or paddling) and mainly wind (using sails – clippers could reach high speeds by using dozens of sails, while transporting cargo), started to combine wind and steam engines and later entirely steam driven later in the 19th century (sailing boats are still popular though, but more for leisure and sport, not so much for transport). Another traditional means of transport was the horse-drawn ship; today the traces of this practice can still be seen in the landscape in the form of tow paths along canals in north-west Europe (the Netherlands, Belgium, France, England, etc.). There are almost no canals in Australia, which is interesting, being a Western nation which was established in this continent around the time that all these changes were taking place. Navigable rivers, inlets, and the ocean were used, but hardly any canals. Particularly in Sydney which offered abundant opportunity for water transport on Sydney Harbour, there was never a canal made between Sydney Harbour as sea entrance (Port Jackson) and the main port in Botany Bay in the south (the Alexandria Canal was never finished, and other 'canals' were mainly open sewages and later just as stormwater drains – currently one of these is being upgraded and restored to a kind of gentrified nature habitat of the creek and tidal areas – though through quite heavy handed machine power, 'naturalisation' by industrial means).

Steam engines were becoming efficient enough to be used on trains and boats, which was a big development, as David Landes discusses, and fuel efficiency was important in this case as every bit of space needed for coal was lost to cargo.[56] The invention of the steam turbine engine (the turbine principle was already in use in water-driven structures) made the machines more efficient. Later, internal combustion engines started to be used in boats (mostly diesel), trucks, and cars, and currently there is a transition to electric cars (though electric boats are still rare).

Electric trams for urban transport of people have been common since the 19th century, also underground in tunnels, subway, or metro trains. Cargo and passenger long-distance trains applied diesel engines, and later electric locomotion.

Road transport changed as well, from horse-drawn cars to the automobile – a car with an internal combustion engine for kinetic energy. Although there were earlier experiments with steam-driven cars, and even electric cars, the internal combustion engine became the most common until today. Petroleum-based fuel delivered a yield in kinetic energy from the stored fuel long unrivalled. Only in the last decade with the improvements in battery technology for the storage of electric charge has made electric cars more feasible.

It is interesting to note that only after more than a century of electric rail transport (trams and trains, and the occasional surviving electric 'trolley' buses in cities like Arnhem, the Netherlands, and Salzburg, Austria) there is a serious development in electric cars. The interaction of driving an electric car is different from a traditional car, as the electromotor delivers the highest torque at low revolutions of the engine, as opposed to the internal combustion engine which delivers peak power at high revolutions.

Manufacturing developments

In the early 20th century, mass production techniques were further developed, leading to another phase of industrialisation (some call *this* the Second Industrial Revolution – further discussed below). The approach of interchangeable parts was developed, which meant that different types of products could be manufactured using standardised components, and this modularity in a certain way could also be used for the purpose of customisation. An important development was the full-scale introduction of the assembly line, particularly in the automotive industry (the T-Ford as the main example). This further reduced the role of the human to being a servant of the machine-based systems. Although commercially successful and very practical (and efficient), it is actually quite an absurd situation to have people following the machine rather than the other way around. The division of labour, which as discussed started in the earlier historical stage of settlement around 11,000 years ago, was developed to an extreme during this industrial stage. The connection between people and nature was further distorted, and replaced by a disconnected and fragmented situation. The notion of specialism was taken to an extreme; people became part of the machine, and very small parts indeed.*

---

* In this situation, people acted as mere cogs in the machine, as depicted in classic movies from the early 20th century such as Fritz Lang's *Metropolis* in 1927 and Charlie Chaplin's *Modern Times* in 1936. But there were also more optimistic movies, less narrative and more focussed on patterns and rhythm and the positive aspects of technological advances, such as *Berlin, die Sinfonie der Großstadt* from 1927 and particularly Dziga Vertov's *Man with a Movie Camera* from 1929, depicting the progress of the industrialised world illustrating the same key themes of manufacture, transport, and communication. *Man with a Movie Camera* is a silent movie to be accompanied by unspecified live music, which is still popular with contemporary musicians. Tom Cora (cellist and composer) played to it many times in the 1990s, emphasising the tension between improvised music performance and prerecorded visuals as I discuss elsewhere [Bongers 2006, p222], Michael Nyman has composed for it, and the Cinematic Orchestra created music for it which was released on DVD in 2003.

Many techniques for mass production were developed in this time, such as injection moulding which allowed the manufacture of large quantities of identical parts and products in plastic and other materials at a low cost. Eventually, this led to the realisation that, rather than the industry merely responding to an existing demand, it could actively participate in creating the demand artificially in the consumerist model.[57] While the short-term economic benefits are clear, the long-term unsustainability of this model, and its mid-term negative effects on the environment, were ignored for a long time (particularly by those who profited from it).

Artistic responses to industrialised manufacture

There were several responses in society to the changing role of the crafts-person in the industrialised manufacturing systems.

In visual art and music, Romanticism was a response seen by many as a critical attitude towards the changes, with new artistic expression with a focus on nature and aesthetics instead of the rational and mechanistic. This was most prominent in the first half of the 19th century.

The *Arts and Crafts Movement* of the late 19th century rallied against the demise of the artisan, noting that the craft of skilled work was replaced by machines. Most prominently the designer and printer William Morris and the theorist John Ruskin were involved in driving this anti-industrial movement, which advocated the return to the artisan culture of "a single person conceiving and making an object from start to finish".[58] They actively developed strategies aiming to find a better balance between craft and industry, through the notion of *Design* – the act of achieving planned outcomes through the involvement of the craftsperson in the *process* of manufacturing. This shifts the emphasis from making an object to preparing an industrial process to manufacture the desired object. This is an important step in the development of industrialised manufacture.[59]

In the early 20th century, the *Bauhaus* school in Germany, in Weimar (from 1919), later in Dessau (from 1925), and for a very short time in Berlin (1932–1933), developed the notion of Industrial Design and modernist architecture. While supporting the notion of the craft in skilled work of the artisan they also saw the role of craft at the beginning of the design process. The idea is that a design made by a skilled craftsperson can be manufactured by machines. This change in role, from craft to design, was worked out by the Bauhaus staff and students, led by the directors, the architects Walter Gropius and later Ludwig Mies van der Rohe. Their approach to design was already common in architecture, where the designers create the plan (the blueprint, the 'design') which is then carried out by builders, artisans, and machines. This is similar to musical practice, where for centuries it has been common for composers to 'write' their musical ideas and structures in elaborate scores (partitures) and the musical notations to be performed (with varying degrees of interpretation allowed) by skilled musicians.

The Bauhaus approach brought together the notion of design with art and technology, through many cross-overs and interdisciplinary explorations. For instance colour was researched and analysed by the painters Josef Albers and Johannes Itten, the painter Wassily Kandinsky carried out his experiments on the relationship between music and abstract visual art, László Moholy-Nagy worked with chemicals and light, and created kinetic mechanical sculptures (which didn't survive entirely, but the Van Abbemuseum in Eindhoven in the Netherlands has a restored version of one of the sculptures), the painter and sculptor Oskar Schlemmer fostered cross-overs to performance and dance with the costumes for dancers in the Triadic Ballet and mechanical performance in the Mechanical Ballet (a reproduction of this installation exists), and in Paul Klee's visual work the influence of his musicianship was apparent.

It was also a period of great political turmoil in Germany, in transition from separate states to one country, between two world wars, all taking place under the influence of, or opposing to, ideologies as diverse as fascism, communism, and socialism. This situation was reflected in the Bauhaus developments in various ways. When the Nazis finally managed to close it down in 1933, many artists and designers involved moved abroad, most prominently to the US. There, they had a particularly strong influence by teaching and setting up new degrees at various universities: Walter Gropius at Harvard, Moholy-Nagy founded the *New Bauhaus* in Chicago, Mies van der Rohe also in Chicago at what is now the Illinois Institute of Technology, and Albers at the Black Mountain College.[60]

PHASES OF INDUSTRIALISATION – THE UNINDUSTRIAL REVOLUTION

A clear result of industrialisation, accelerated by wartime endeavours, is the level of mass production reached by the mid-20th century. And as mentioned before, at the same time the 'information age' was developing, with mass media and many new ways of communicating – first through analogue electronic media, later digital. In the next chapter, information representation, semiotics as the study of signs, meaning making, and communication are discussed in more detail.

Sometimes this combination of mass manufacturing and information technology is referred to as the *Second* Industrial Revolution, and this is getting a bit confusing, this would be the third definition of the Second Industrial Revolution discussed in this chapter (in the overview below, phases 2, 3, and 4). Furthermore, it is often said that we are currently in yet another phase of industrialisation, of manufacturing under the influence of distributed computing, robots, artificial intelligence, and new manufacturing techniques such as 3D printing. A special report in the *Economist* magazine in 2012 covered this under the header of 'The Third Industrial

Revolution',[61] which is also the title of a best-selling book by the economist Jeff Rifkin published in 2011. To add to the confusion, it is currently also popular to refer to these developments as 'Industry 4.0', or *The Fourth Industrial Revolution*, the title of a book by Klaus Schwab of the World Economic Forum published in 2017, covering more or less the same amalgamate of not necessarily related developments.

Of course the naming of the industrialisation phases depends on the emphasis; in the context of this book the focus is how relationships and interaction between people and manufacturing technology have changed over the developments. It is through this lens, but also taking into account the broader impact of industrialised manufacturing on society, that a number of key developmental phases have been identified in this chapter, summarised below:

Industrialisation Phase 1) the first industrialisation was about the shift from manual, craft-based labour to machine-based manufacturing, driven by coal (second half of the 18th century).

Industrialisation Phase 2) the next phase of industrialisation was about the combination of machines with internal combustion engines, with electrical technologies (from the late 19th century, which for instance David Landes calls the Second Industrial Revolution indeed[62]).

Industrialisation Phase 3) early in the 20th century, another important phase was about the standardisation of components and the streamlining of manufacturing through the assembly line. This led to many productivity gains as well as further influencing the relationship between the workers and the machines – mostly to the detriment of the former). This is (also) often called the Second Industrial Revolution.

Industrialisation Phase 4) this phase is about the combination of manufacturing and information technologies, from the mid-20th century.

Industrialisation Phase 5) the increased use of robots in the manufacturing process, as complex and versatile participating machines, in the second half of the 20th century.

Industrialisation Phase 6) since the early 21st century, the current phase which is about manufacturing in combination with robots and AI, distributed computing (IoT), AR&VR, and 3D printing; all the contemporary buzzwords – but particularly 'instant manufacturing'.

Rather than arguing for labelling this last phase (which already has been labelled as the third, and the fourth) as the 'Sixth Industrial Revolution', I prefer to call it the *un*industrial revolution. It is about the shift away from mass manufacturing and industrial structures, enabling manufacturing on an individual, customised, and personalised basis, not mass produced. Of all the current influences often mentioned such as IoT, AI, and 3D printing, it is particularly the influence of the technology of 'computer-modelling-based instant manufacturing' on industrialisation which has the biggest impact on society. The other two can be seen as extensions and developments of earlier phases, so the role of IoT or distributed computing can be linked to Industrialisation Phase 4) from the list above (combination of manufacturing and information technologies) – and AI can be seen as an extension of Industrialisation Phase 5) in the list (the use of robots in the production process). The crucial differentiating developments in this sixth phase are the computer-based production techniques, such as 3D printing, CRC milling, Laser Cutting, etc. All these machines produce physical artefacts based on computer models (traditionally called CAD/CAM: computer-aided design/manufacturing). Particularly 3D printers are popular to make objects directly, without a factory, often used for making prototypes to try out products ('rapid prototyping') but also for manufacturing.[63]

The discerning feature of these technologies is the possibility of making unique products, instead of mass-produced identical products; individual production instead of mass production. Of course, although the 3D printing machines are getting faster and more precise (such as the high-end SLS machines), the creation of each artefact in this mode of production is much slower than in mass production (such as injection moulding) – but the whole process from idea to artefact is much faster. Setting up a traditional industrial mass production process is slow, tools (the moulds) need to be produced, there is little flexibility (only achieved through modularity – Lego is an extreme example), and the on-costs and investments are relatively big which means risk and exploration will be avoided. I therefore prefer the term Instant Manufacturing for this technique, as it leads to the possibilities of individual products (or 'mass-customisation' as John Heskett called it [2002]). Instant Manufacturing is a much more agile process, with flexibility and room for exploration, and opportunities for personalised designs. This notion has also been developed on the architectural scale, using a 'file-to-factory' technique to create buildings not based on standardised material shapes but individually produced elements, as convincingly demonstrated by the architect Kas Oosterhuis in several of his buildings.[64]

The implications and future possibilities of instant manufacturing are further discussed in Chapter 12 *Understanding*, with examples of projects that I have been involved in, developing individual devices as 'products on demand' with the potential to customise to the person's specific needs.

CONSEQUENCES OF INDUSTRIALISATION

There is a sense of accelerating change; where the first development (of tools, language, culture, and the human mind) took several millions of years but particularly the last few hundreds of thousands of years, the second key development of agriculture and settlement (villages, cities, urban culture) took several thousands of years, and the third development (Enlightenment, scientific and rational methods, Industrialisation) took several hundreds of years. The recent developments seem to have happened so fast that we can barely adjust to them! Yet we need to understand them in order to design responses and apply technologies appropriately.

When studying technological developments, it is often noted that change begets change – each development leads to and supports further developments.[65] Due to this incremental and exponential nature of technological change, developments easily spiral out of control, often with unforeseen and unanticipated (and not necessarily positive) consequences.

In this section some of the negative consequences of industrialisation are discussed, such as the role of industrialised warfare, and some of the adverse effects on the environment as a result of industrialisation, to show why it is important to be more aware of these developments.

Warfare and industrialisation

In the context of industrialisation, warfare can be seen as an extreme of an industrial (machine-oriented) system. The First World War was approached in this extreme, with soldiers in trenches instead of factories, with a huge loss of human life. Lewis Mumford discusses the relationship, by comparing the mines with the trenches in *Technics and Civilisation*[66] and also noting the increased demand for iron which was driven by the weapons industry. Writing about the First World War, Mumford stated that "it was a large-scale industrial operation" also with a "meaning in reverse: modern industrialisation may equally well be termed a large-scale military operation". Indeed, compared to the trenches of industrialised warfare, people felt they were better off in the factory.[67]

The subsequent wars also claimed massive amounts of human lives, often as 'collateral damage' in the attempts to achieve the destruction of military equipment. The physicist and mathematician Freeman Dyson, reflecting on his involvement in the planning of bombing raids of the British on Germany during the Second World War, illustrates how wartime industry and bureaucracy was operating as an unstoppable machine, continuing their work in destructing and taking human lives when it was clear that it had no benefit for the eventual ending of the war. This was during an evening-long interview on Dutch television in 1993.[68] In his memoirs, *Disturbing the Universe* [1979], Dyson calls it "a bureaucratic accounting system which failed utterly to distinguish between ends and means".[69] He further discusses a number of irrational situations related to the ergonomic factors of the bombing planes, shortcomings in the design (trivial issues such as the size of the escape hatch being too small for some of the crew to fit through), and training issues, crucially related to potential survival of the crew, which were not allowed to be discussed and took years to address.[70] Perhaps the most shocking are the repeated revelations about how inefficient the whole endeavour was, and how pointless in the bigger picture of deciding the outcome of the war. The exception to the inefficiency was the 'success' of the tragic firestorm in Dresden in the east of Germany, which had however very little strategic value. Dyson also presents some useful critical insights in the notion of 'deterrence'. In the TV interview he makes a connection between the Second World War and the Gulf War, which had just happened in the years before the interview, with the fallacy of 'precision bombings'.[71]

Focussing on the bureaucratic structures manifest in warfare throughout history explains some of their profound inefficiency, meaninglessness, and insanity. Emphasising the role of electronic transmission of orders, Marshall McLuhan discusses this delegation of authority in Nazi Germany (first in his essay 'Culture Without Literacy' [1953], discussing the means of communication, and later in an early version of *The Medium is the Message* [1960] in the context of a discussion about the effects of decentralisation in industry and corporate structures). McLuhan refers to a speech by Albert Speer at the Nuremberg Trials after the Second World War in which the Nazi architect reflects on how the whole notion of 'following orders' changed with electronic media, with Speer stating that

> Former dictatorships needed collaborators of high quality even in the lower levels of leadership, men who could think and act independently. In the era of modern technique an authoritarian system can do without this. The means of communication alone permit it to mechanize the work of subordinate leadership. As a consequence a new type develops: the uncritical recipient of orders.

Indeed, Albert Speer is known not just as the architect of many Third Reich buildings of the Nazi Germany era, but also the one in charge of successfully streamlining its industry for warfare (pitching this against the British machine as described by Freeman Dyson mentioned above) – and himself indeed in a dubious position (with regards to responsibility) as an 'uncritical recipient of orders'.

This top-down structure is in sharp contrast with the contemporary situation of terrorism, which operates in a decentralised manner, effectuating terror in a localised way. In fact this is the strength of terrorism: its independence, being decentralised, distributed, very much like the basic structure of the Internet – ironically devised indeed on purpose to be decentralised, as an ultimate strength to deal with calamities. This approach is also fundamental to the working of Bitcoin, as opposed to the centralised structure of all other currencies. By its very nature a decentralised system is much more robust and less sensitive for disturbances, unlike a centralised system as described above.

Generally, bureaucracy is eagerly supported by technology, as can be seen in one of the earliest examples of the Roman Empire where those in power used papyrus to extend the reach of their commands over far greater distances than previously possible. Marshall McLuhan reflected on this (discussing it in the context of money, that symbolic representation of value), which gave the Romans a competitive advantage over the Phoenicians who used clay tablets which were much heavier – as a sea faring nation, this initially was not an issue for the Phoenicians.[72]

Human nature

It can also be argued that the notion of the 'uncritical recipient of orders' does not necessarily resonate with our 'human nature' as is often thought – in fact, we are more inclined to do good than to do bad, as historian Rutger Bregman convincingly asserts in his recent book *Humankind – A Hopeful History*.[73] His argument could be seen as too optimistic (he is a positive thinker, as shown in his previous book, *Utopia for Realists*[74]), but the key to Bregman's argument is that what is seen as 'human nature' since biblical times (Adam and Eve eating that apple, Pandora opening that jar) and millennia of warfare and trouble, that this negative view is rather a by-product of top-down, centralised, and hierarchical structures in our civilisations. Jared Diamond has many examples of humankind's unkindness in his books, and Bregman's earlier book too (*De Geschiedenis van de Vooruitgang*, 'the history of progress' mentioned above, and also *Utopia for Realists* restricts the view to recent times). What made Bregman more nuanced is that for his *Humankind* book he went back to the 'prehistoric' times and treated with more insight and respect the way we lived and developed technology and culture over millions of years (as discussed at the end of Chapter 4 *Settling*). In *Humankind* he presents his further research into debunking a number of myths and commonly accepted stories about the negative view of 'human nature'. For instance William Golding's famous novel *The Lord of the Flies* from 1951, and the 1963 movie based on the book, about a group of boys that have to survive on a deserted island and turn to savagery – but as compelling as the story seemed, it always was just that, a made-up story. It seemed particularly plausible from Golding's context of British society in the 1950s, his own personality, and the general view on human nature at that time. The story has been popular ever since, and many other stories, films, and TV series (such as *Lost* and *Squid Game*) explore similar themes.

However, Bregman tracked down the survivors of a real case, of a group of six Tongan boys who were shipwrecked on a desert island in June 1965 – and managed to thrive for more than a year in cooperative harmony. Bregman interviewed the Australian captain of the boat who rescued the boys, as well as one of the survivors, who are now both living near Lismore in northern New South Wales.[75] From these interviews and other sources Bregman studied, a clear picture of peaceful development and successful survival emerged.

In *Humankind*, he presents how a range of social science experiments from the 1960s which reported participants' willingness to be cruel to each other turn out to be misinterpreted too, for instance the common story of the Milgram experiment, in which unwitting participants were compliant in administering electric shocks (to harmful and potentially lethal levels, to a maximum of 450 volts) to other subjects. The range of experiments were devised by social psychologist Stanley Milgram in the early 1960s at Yale University, who wanted to prove that innocent and decent people would comply with following orders and carry out cruel tasks as 'uncritical recipient of orders'. These and other similar experiments were directly related to the aftermath of the Second World War and particularly the Holocaust, trying to understand why people went this far in such terrible behaviour. Even ordinary people, it seemed, were involved in this during the war −"the banality of evil" as philosopher Hannah Arendt famously put it at the time. To study this, the Milgram experiment used a set-up with an electroshock machine that was not real, but the participants did not know that − it was an effective piece of theatre (as Milgram actually liked to put it, he was inspired by the Candid Camera shows on TV at the time). The results as reported at the time (and repeated ever since) suggested that a shockingly high number of the participants seemed willing to carry out the tasks in some of the over twenty different versions Milgram created (and not all reported), as high as 65%, following orders even if this seemed to lead to bodily harm to others. This is in sharp contrast with estimates of experts and laypeople at the time, which usually ranged in a few percent. Needless to say, an experiment like this would be totally unacceptable by contemporary academic research ethics standards, if only because of the element of deception in the set-up, and the many improvised elements as already pointed out at the time as a flawed methodology.[76] We now know that many of the in total over 1000 participants, some of which were strongly traumatised by the experience and had feelings of guilt and remorse, were not briefed about the true nature of the experiment (they were told it was about memory and learning) until much later, if at all. Bregman discusses recent research into the original transcripts and data of the experiments, by among others Melbourne academic Gina Perry who wrote the book *Behind the Shock Machine − the untold story of the notorious Milgram psychology experiments* [2013], which shows a much more nuanced picture. It turns out that in fact the majority of the participants were openly reluctant in administering the 'shocks', resisting, arguing, rebelling, and even downright refusing to carry out the task. And those who did were actively pressured or even bullied into submission. There were also a significant number of participants who doubted the experiment (and the electric shocks) were real, which made it easier for them to comply as they realised the shocks were fake. In the end, Bregman concludes, all that can be inferred from these experiments is not the urge to be obedient, but to be compliant.* This human urge to comply is very strong, particularly for young children, but also for adults given the right context. The Milgram experiment was set up to exploit this tendency, staging a situation that had all the urgency of a scientific experiment which strongly relied on the participants' efforts, which was made very clear to them. Rather than 'uncritical recipients of orders', many situations and structures in which the worst of human behaviour comes out are based on the exceptions (not the norm) of psychopathic characters, and as we now know many of the main players in Nazi Germany acted as they did not because they were told to in great detail, but because they thought it was the right thing, they were actively trying to impress their leader (der Führer),[77] as a 'conformity with evil'. Milgram's experiments show that in order to achieve conformity, "evil has to be disguised as doing good" as Bregman puts it.[78]

In other chapters Bregman discusses human behaviour in battles during wars, and presents the evidence of soldiers actually deliberately trying to avoid killing or even harming the enemy. This explains some of the deficiencies in the effectivity of the Allied bombing raids on Germany at the end of the Second World War mentioned above, and discussed by Freeman Dyson. This is very different from how things are depicted in the many war movies and books that feed the common imagination (one exception is *The Thin Red Line* by Terence Malick in 1998). Similarly many post-apocalyptic and disaster movies and stories, such *The Road* (from 2009, based on the 2006 novel by Cormac McCarthy) or *War of the Worlds* (based on the 1898 novel by H. G. Wells) and particularly *Mad Max* (by George Miller since 1979), routinely show how survivors act selfishly, and behave with hostility to each other, again in stark contrast to what has been witnessed in real-world situations. Fictional narratives are driven by the same potentially deceptive self-image, as self-fulfilling prophecies, and it is important to keep a focus on empirical studies and actual events. In an interview in May 2020, *Black Mirror* author Charlie Brooker reflects on how he noticed the great amount of care and acts of kindness that people showed for each

---

* The film *The Experimenter: The Stanley Milgram Story* from 2015, directed by Michael Almereyda, gives a good insight into Milgram's character and fascinating experiments, and although very well made with some great acting and sometimes intriguing cinematography, the film doesn't mention the recent and earlier critical interpretations of the research [Nicholson 2016]. Milgram himself made a 44-minute-long documentary (with the help of a filmmaker) in 1965 of one of the experiments that he carried out in 1962 [Perry 2013, pp281–287] (fragments have been broadcasted on TV, and on YouTube, the whole documentary is available through Alexander Street Press). A TV film of 1976 shows a dramatized account of Milgram's research with a particular emphasis on the ethical issues of carrying out these experiments [Perry 2013, pp272–274, pp289–290] – not a great film but if only for catching a glimpse of a very young John Travolta who appears (uncredited) as a lab technician behind the glass of the experimental set-up (a low quality copy of the film is on YouTube). Australian academic and film maker Kathryn Millard, who is very critical about the experiments [2015], created a remake of the documentary (this time with real actors) in 2015 called *Shock Room* (which can be found on Kanopy).

other during the COVID-19 pandemic in 2020 (apart from some selfish and unnecessary hoarding of toilet paper). He speculates that he wouldn't be able to write such a dark depiction of human behaviour in future scenarios and post-apocalyptic situations when "what's actually happening at the moment is much more cohesive and heartening".[79] Again, what the top-down structures lacked in compassion (the detrimental and perfidious competition between countries for resources mentioned earlier) seems abundant in bottom-up societies and socially driven responses.

It is obvious that humans are by nature caring and cooperative, having evolved as social animals for hundreds of thousands of years. As Bregman shows in *Humankind*, the strength of our societies, particularly in democracies, is at the same time the weakness: the hierarchic structure and top-down ruling. This can be hijacked by despots and dictators, allowing the tiny minority of narcissists, malicious characters, or even psychopaths to take over – at any level, as we have seen in the US presidency from 2016–2020 – but also in organisations, particularly commercial corporations. There seems to be a tendency in Western and particularly Anglo-Saxon societies to favour the overly (and often misplaced) self-confident, 'strong', alpha-person leader who makes all the promises that people want to hear (from voters to recruiters) even if they are unrealistic, paradoxical, or just false – although some cultures are more merit based. The future vision of elected leaders and parties is usually limited to the horizon of their tenure, often just a couple of years. This is why other structures are so important, such as the international collaborative bodies such as the United Nations, World Health Organisation, and the European Union.

While top-down structured 'civilisations', from settlement to industrialisation and beyond, have shown to be vurnerable to these pressures, in traditional societies such as hunter-gatherer cultures, which are based on consensus and cooperation, such behaviour and character traits is actively discouraged and curbed.[80]. This is discussed in the next section.

Humans and nature

A good example of a consensus and mutual respect based society is the traditional Australian Aboriginal culture. I am cautious with this, as I am aware of the many cultural tensions and sensitivities around the issues in the Australian continent, and there is always the risk of romanticising the virtues of Indigenous cultures, or worse, misrepresenting ideas. So, I am avoiding any value judgement, just discussing what I think we can learn from, and be inspired by in, other cultures, as mentioned in other parts of this book (towards the end of Chapter 4, and later in Chapter 6). In the many sources, and in my own experience living here for over a decade, including several close connections I have been lucky enough to have been involved in, it is clear that the phrase which is often used of the 'original custodians of the land' is an apt one, and it is the crux of exactly what Western society at large is missing. As discussed in previous chapters, people have been in the Australian continent for over 50,000 years, and in that period the traditional Australian Aboriginal culture developed an elaborate and appropriate relationship with their environment. This is in stark contrast to the Western capitalist model with its inherent emphasis on eternal growth, which is an obvious fallacy. In the traditional culture a balance was found between the number of people and distribution, living as much as possible in harmony with the land ('on country'). This included the practice of 'fire-stick farming' as discussed in Chapter 4, which not only served the immediate needs of hunting but also supported the long-term benefits for the local flora. There are many examples: one is the practice of leaving middens (shells and other remains of maritime feeding activities, still present in the current landscape), which had the purpose of communicating to future inhabitants of the area which food was taken from the environment in order to be able to protect it from being overexploited in activities such as overfishing. In the traditional myths, beliefs, and culture, there is an intricate set of relationships and interdependencies between people and environment. Of course this was the result of tens of thousands of years of development, and there is evidence that some big changes in flora and fauna occurred around the time the first humans reached the continent, such as the disappearance of large marsupials or megafauna which are always the most vulnerable (it happened everywhere around the world where *Homo sapiens* appeared).[81] Although the cause is debated, it does seem likely that the arrival of humans had an influence on the delicate ecology which was previously isolated for millions of years, but the recent Ice Age played a role too.[82] And anyway it pales in comparison to the large-scale destruction and extinctions that have taken place since 1788 with the arrival of the first British colonists, or as biologist Tim Flannery put it in *The Future Eaters* [1994], "The arrival of a new people, with no understanding of the ecology of their new homeland, was to prove the undoing of 35000 years of conservation effort." [83] This included the introduction (accidentally or on purpose) of non-indigenous species of animals, plants, and trees, with far-reaching negative ecological consequences. The traditional Aboriginal culture has since been severely damaged, trampled, and suppressed for about two centuries, but in recent decades it has become increasingly appreciated. Many attempts to restore the traditional culture, and general cultural reconciliation, have taken place, at least by some. Traditional Aboriginal culture is finally seen as a source of inspiration and learning about alternative and often much more appropriate approaches to life and the relationship with our environment – firmly supporting the importance of the notion of custodianship or stewardship. The challenge for humankind at present is to adapt and extend this notion of custodianship, and apply it to the whole planet. The ecosystem we are part of, the whole complex system and self regulating of planet Earth that James Lovelock has called Gaia since the 1960s (as mentioned at the end of Chapter 3),[84] would benefit from such an attitude. Gaia is the name of the Earth goddess in ancient Greek mythology (and ironically it was William Golding, of *The Lord of the Flies*, who suggested this name to Lovelock)[85].

This interpretation of the whole of Earth's ecosystem as an organism, Gaia, in a way is a metaphor at least, which can be very useful as an approach to find a better balance between people, nature, and technology.

Humans *versus* nature

Currently, our new powers through technology and science seem to have made humankind think (and act like) they possess 'godlike powers', as is the central thesis in Yuval Noah Harari's second book *Homo Deus* [2016]. This same notion was discussed earlier (and more critically) by Lewis Mumford in his book *Technics and Civilisation* [1934]. Mumford wrote "man could achieve godhood"; however, he observed that this was done by "renouncing a large part of humanity", and actually warned that humankind "created the machine in his own image: the image of power, but power ripped loose from his flesh and isolated from his humanity".[86] In this light, humankind's desire was to dominate, rather than cultivate,[87] which is a sharp contrast to the Australian Aboriginal attitudes discussed at the end of Chapter 4 and in the previous section. Mumford was intrigued (and somewhat horrified) by the potential of "the machine". He realised that by rationalising the environment, humankind by necessity had downplayed many subtleties and actually many essential factors of reality. As he puts it: "By his consistent metaphysical principles and his factual method of research, the physical scientist *denuded* the world of natural and organic objects and turned his back upon real experience."[88] In other words, what we made, the machines we created, suited this world view, *not* the real world.* And while

---

* This notion that the active mechanical technology increasingly detaches people from their environments is clearly (and painfully) illustrated in the case of industrialised agriculture by John Steinbeck in his 1939 novel *The Grapes of Wrath* written in (and about) the same era of the Depression. In Steinbeck's novel, he describes the disconnection so aptly (on page 38) that I quote it here in full: "Behind the tractor rolled the shining disks, cutting the earth with blades – not ploughing but surgery, pushing the cut earth to the right where the second row of disks cut it and pushed it to the left; slicing blades shining, polished by the earth cut. And pulled behind the disks, the harrows combing with iron teeth so that the little clods broke up and the earth lay smooth. Behind the harrows, the long seeders – twelve curved penes erected in the foundry, orgasms set by gears, raping methodologically, raping without passion. The driver sat in his iron seat and he was proud of the straight lines he did not will, proud of the tractor he did not own or love, proud of the power he could not control. And when that crop grew, and was harvested, no man had crumbled a hot clod in his fingers and let the earth sift past his fingertips. No man had touched the seed, or lusted for the growth. Men ate what they had not raised, had no connection with the bread. The land bore under iron, and under iron gradually died; for it was not loved or hated, it had no prayers or curses." Earlier in the chapter, the farmer, about to be driven from the land due to its unproductivity caused by drought, explains how this is a result of the monoculture (of the cotton plantation) by necessity due to the lack of funds, which would be further exacerbated by the industrial approach to farming. This is decades before the importance of crop rotation was fully realised, and even further away from the notion of permaculture. The resulting ecological disaster at the time, known as the 'Dust Bowl', drove the colonists to the west, which is the main topic of the novel. Steinbeck in Chapter 5 describes how the farmers are evicted from their land by the owners – the banks – and presents banks as faceless, unhuman entities, "machines", in the same way that Mumford presents the 'mega-machine'.

Mumford wrote all this during the Great Depression era, it was still before the Second World War, and before nuclear power.

It seems that this detachment of experience and lack of respect is still prevalent; all further technological developments seemed subservient to this goal of domination – but with its increased sophistication, contemporary technology (digital, interactive, interconnected) offers the opportunity to rectify this misfit (this is further discussed in Chapter 12 *Understanding*).

While, as discussed above, since the 16th century we started to accept ignorance, and learned how to deal with it … but it seems that we still haven't accepted *uncertainty*. When looking at scientific developments critically, it is clear that in spite of all their virtues and benefits, the approach does seem to have a tendency to interpret, translate, and ultimately *reduce* the complexity of the natural world to such a level that it can be managed. As Danish science writer Tor Nørretranders put it, science is about being unambiguous.[89] Reductionism's potentially fatal tendency to dumb down extends most prominently in science's cousin, engineering. This means, as Lewis Mumford already emphasised in *Technics and Civilisation*, that technology is developed based on a limited, reduced ("denuded") interpretation of reality. Our technology, therefore, in spite of all its obvious virtues and enabling power, is often inherently flawed and suboptimal. In the last decades at least, technological developments show signs of maturing, and many thinkers in the field that studies the design of interactive technologies have advocated and actively contributed to a more appropriate form of technology – for instance Mark Weiser from Xerox PARC (Palo Alto Research Centre) in the early 1990s famously discussed the notion of *calm technology*.[90] Meanwhile, from Frankenstein's monster in Mary Shelley's novel (first published in 1818) to *The Matrix* movies by the Wachowskis (starting in 1999), the notion that machines will attempt to take over humankind is a recurring theme – usually with suboptimal consequences for the latter. Yuval Noah Harari actually emphasises a future scenario where humans will become the pets of the next generation of evolution, the machines.[91] This may not be that bad actually, for those currently indulging in hedonistic lifestyles there may not be much of a noticeable difference, but perhaps we can aspire to some more ambitious attitudes to the world and actually living up to (what could be seen as) our responsibilities. James Lovelock writes about this too in his book from 2019 *Novacene – the coming age of hyperintelligence*, mentioning this notion of pets or even plants, kept by our successors in the 'novacene'. He calls them cyborgs, the human-made machines and robots that become cleverer and more powerful than us, which is known as the 'singularity'. (Cyborgs are generally defined as a mix of organic and artificial life though.) Lovelock sees in a role of parent and midwife to this new form of life that we are an essential facilitator (like a catalyst in chemical processes, by the way) of this next step in evolution.[92] I see it this way too, as discussed in more detail towards the end of this book in Chapter 12 *Understanding*. I am actually a bit more critical as I think we should focus on the current stage which is about humans, nature, and technology in symbiotic relationships and not (yet) about technology taking over. Linear and unimaginative extrapolations based on numbers are not convincing. Throughout his book, Lovelock emphasises the importance of our powerful ability of intuitive thinking in addition to rational reasoning, which is further explored in Chapter 10, *Thinking*, which discusses the relationships between rational and unconscious thought processes.

While acknowledging the tremendously important contributions that science has made to the advancement of knowledge, it is also important to be aware that, after the initial phase of great discoveries and breakthroughs (such as the insights by Newton on gravity) which inevitably in a historical retrospective view have a certain sense of opportunism and dash for the 'low hanging fruit', science in some instances has gotten a bit bogged down in the complexities of reality and therefore perhaps started to attempt to reduce the complexity.[93]

There was this joke I remembered from my childhood. (It may seem totally inappropriate and unacademic to present jokes, but there is often a lot of wisdom encapsulated in these little narratives.) It is a well-known joke, which goes something like this: it is about a drunkard, who lost his keys to his house late at night and so he couldn't get in. He was searching for his keys under a street light, and when he got interrogated by a police officer about his endeavours at this particular location he admitted that he actually didn't lose his keys right there, but as he said, "this is where the light is". Much later, as often with such jokes I realised there was a lot of life wisdom in the story, and that it was a great metaphor about some inevitable opportunism in the search of knowledge and truth. And indeed, there is now the notion of the "streetlight effect",[94] which is exactly about this tendency of science to look in places where one *can* look, not necessarily the same place as where one *should* look, just because it is inaccessible or just too difficult. This could actually be alright, as long as one knows that one is doing this – I sympathise therefore with the drunkard in the original story, because at least he knew! This will be discussed in Chapter 12, in the context of a design approach which does exactly this – to be deliberately opportunistic and look at the wrong place – whilst acknowledging it, and by being aware of this transition it is actually possible to infer interesting and useful insights from research. But at this point it is clear that there may be a bit of a problem with some of the scientific research, although not to the extent that David Freedman has it, in his book *Wrong – why experts keep failing us, and how to know when not to trust them* [2010]: he goes a bit over the top by denouncing the role and validity of experts, and actually unintentionally (I hope) supports the whole fake news and new untruth paradigm. But there is a point here, that this endeavour which started off nobly and with great intentions, of admitting ignorance and conquering this wide open field, successfully questioning dogmas and assumptions, perhaps got a bit stuck, watered down to something not as relevant as it could (should) have been. Again, this is very understandable and can actually be ok, if it gets acknowledged. And this is where the tension is. There is a tendency in scientific research, and by extension in engineering approaches to developing technology, to conveniently convene around that streetlight (or the light of the smartphone screen) without being aware of the fact that we are doing this.

It is good to bear this in mind when we look at this 'age of reason', to see it as an essential stage in the development of human thinking and being, yet not necessarily as the final stage. Mumford's stance may seem somewhat romantic, but unfortunately still very appropriate when discussing current developments in the interaction between people and technology. Mumford in the late 1960s wrote about the "mega-machine", which included not just technology but social structures, such as the building of pyramids which emphasise the role of bureaucracy in such developments.[95] This is still relevant, as reflected in a way by communications theorist and NYU Professor Neil Postman in his book *Technopoly – the surrender of culture to technology* [1992] who discerns three phases: *tool-using cultures* (corresponding roughly with the prehistory and settlement phases described above), the *technocracy* (corresponding with technology-based society since the industrial age), and what he identifies as an age where technology dominates culture, the *technopoly* – which is the most recent phase. Bureaucracy, always an extension of certain aspects of the human need to organise and control, is increasingly a dominant force in society. But bureaucracy is a means, one that is however persistently and increasingly being turned into an end. Bureaucratic organisation has no purpose in itself, in fact it can (and often does) obstruct development, inhibit innovation, and stifle creativity. Technology, then, tends to be the ultimate servile extension of the bureaucratic tendencies, through which it can reach unforeseen and often unwanted effects.

The Anthropocene

Since the start of industrialisation many irreversible changes to the face of the earth and impact on other species have taken place. These changes and damage were not always necessary, and they were mostly driven by the attitude of humankind's assumed superiority and claimed rights to dominate the environment and exploit it for our own needs and desires often without the necessary degree of constraint.

In the last decades the term Anthropocene has been used, which places our impact on the planet on the geological time scale. This is following the tradition of geological time scale terminology; currently the Holocene epoch of the quaternary geological period. The term was supported in 2000 by atmospheric chemist Paul Crutzen, who was awarded the Nobel prize for his work on studying ozone layer depletion,[96] although the concept is older and even the term Anthropocene was proposed by ecologist Eugene Stoermer in the 1980s.[97]

The use of the term is somewhat disputed, as geologists usually work on much longer time scales that can only be identified in hindsight when the strata have formed, as studied by a branch of geology called stratigraphy.[98] But there are many geological markers that indicate a significant influence, and an Anthropocene Working Group led by Jan Zalasiewicz has been working on a proposal for the International Commission on Stratigraphy (ICS). They presented their findings in a recent article in *Science*,[99] discussing a number of markers that could be significant in a stratigraphical sense. Particularly the onset of nuclear testing in the 1950s shows clearly in the environment, which led them to propose this as the starting date of the Anthropocene. The starting point is debatable and as discussed in this paper, with its focus on stratigraphical signatures caused by human interference, others would argue that the start of industrialisation is the beginning of the Anthropocene, while some would argue that it started with farming – when the settlement period gained prominence, from about 12,000 years ago (land clearing and agriculture had an impact on CO_2 emissions), or definitely in the antiquity 2000 years ago when these developments gathered a critical mass of influence. Our species has always had a stronger impact on the environment than any other species, considering our size and limited physical resources, ever since we started to use tools and language. Our negative impact has certainly accelerated since industrialisation, and further driven by the somewhat blind technocratic and capitalist profit-focussed developments in the 20th century.

An example of this tendency is the use of lead in fuel for car engines to improve performance, invented in the 1920s. Even though the adverse effect on human health of lead was already well-known, the chemical industry denied this and pushed the application. The invention of the use of tetra-ethyl lead in engine fuel (euphemistically marketed as 'ethyl gasoline') is attributed to Thomas Midgley, an engineer by training, and as Bill Bryson in *A Short History of Nearly Everything* remarks "the world would have been a safer place if he stayed so. Instead, he developed an interest in the industrial application of chemicals."[100]

Midgley is also known for the invention of CFCs (chlorofluorocarbons) in cooling devices such as fridges and air-conditioners, which as we know affected the ozone layer and therefore were regulated and banned in most of the world in the 1990s (shortly after the ban on lead in car engine fuel). Paul Crutzen writes in the Anthropocene article

> Things could have been much worse: the ozone destroying properties of the halogens have been studied since the mid-1970s. If it had turned out that chlorine behaved chemically like bromine, the ozone hole would be then have been a global, year-round phenomenon, not just an event

in the Antarctic spring. More by luck than wisdom, this catastrophic situation did not develop.

While the CFC side effects could perhaps not have been foreseen (but definitely acted upon earlier when it became known, since James Lovelock's discovery using a new device he invented, in 1971[101]), the example of lead in fuel shows the propaganda push that industries are willing to undertake in name of progress and profit (for them, not for the environment, or even other people). Particularly because there was an already known alternative, ethanol, but that was deemed to have less potential for profits. Leaded fuel for cars started to be banned in the 1970s, and since 1985 lead-free fuel became available in the Netherlands, but it wasn't until 30 August 2021 that the end of it was celebrated. Similarly, later examples are the now widely discredited claims of the tobacco industry (the harmful effects were known since 1950),[102] and the still strong fossil-fuel industry, to name a few. There is still a long way to go, in finding a better balance between technology-driven progress for short-term profit and looking after the environment and ensuring long-term prosperity.

We might be the first species proposing to name a geological epoch after ourselves, but does that mean that we are the first species that is creating the circumstances for its own extinction?

A previous occasion of near-extinction took place about 74,000 years ago, with the eruption of the supervolcano Toba on the island Sumatra in Indonesia.[103] This was the largest eruption of the last 2.5 million years, many times stronger than the eruption in December 2018 of Anak Krakatau in the same area, which caused a destructive tsunami in the region. The Toba eruption of 74 kya had a vast impact on the earth's climate and liveability, particularly in the area to the north-west of the event, reaching at least as far as Greenland (volcanic ash was found in drill cores). The ash clouds at the time caused a volcanic winter that lasted for several years or even decades, and it is thought that it brought our species close to extinction, with only about 50,000 people surviving.[104] This explains the narrow genetic base of our species. It is also possible that the fall-out of the event in later years played a role in the first colonisation of the Australian continent, as it is in the less affected area south-east of the eruption.[105]

Throughout history there seem to be a constant ebb and flow of wars between people, and between people and the environment. The war on nature also affects us as a species, but particularly affects the many other species of animals and other organisms, many of which have become extinct or are about to become extinct – at an astonishing rate (even the elusive platypus is under pressure[106]).

Yuval Noah Harari in *Sapiens – a brief history of humankind* relates the three phases of human development (the cognitive, agricultural, and industrial revolutions) with three distinct waves of extinction, "Long before the industrial revolution, *Homo sapiens* held the record among all organisms for driving the most plant and animal species to their extinctions. We have the dubious distinction of being the deadliest species in the annals of biology."[107] And just as the first two stages were related to climate change (the first that led our ancestors to move out of the trees and into the open savannahs, developing tools and language, the second at the end of the last Ice Age which enabled settlement, agriculture, and writing), our third phase is inversely related to climate change – one that we are causing.

Anyhow, this is an indeed dubious honour that we have bestowed on ourselves as a species, but perhaps by realising the dangers of our impact on the environment humankind might realise that with this ability to profoundly damage the planet, we might be equally able to save it. After all, 'sapiens' means 'wise'. I hope that by understanding our interaction with technology, we will know how to develop and design our technological environments further for a better future, resonating with the natural environment in a symbiotic relationship instead of plundering and exploiting it. People who doubt these effects seem to have the optimism of (from another joke, sorry) the person who fell from of the top floor of the skyscraper, and on the way down while falling past something like the 15th floor, confidently proclaimed: "so far so good!"

NOTES

1. Sir Ernst Gombrich's classic *The Story of Art* was first published in 1950, and has since been reprinted with additional introduction texts. I have read the 16th edition from 1995, which is the edition that is referenced here [1995, p223].
2. [Bregman 2013, p153].
3. [Gombrich 1995, p223].
4. Harari writes about this in *Sapiens* [2014, pp275–306, Chapter 14], the quotes are on p279.
5. The phrase is also well known in Latin, *cogito ergo sum*, and in French *je pense, donc je suis*, as Descartes' own escape hatch of 'le doute cartésian'. Rutger Bregman discusses Descartes and the doubt ('twijfel' in Dutch) in his second book, *De Geschiedenis van de Vooruitgang* [2013, p157] ('the history of progress', which is only available in Dutch, first published in 2013; I used the 2019 edition which seems the same as the original but with a new preamble). Harari doesn't mention Descartes, which perhaps is quite refreshing. Generally, Harari focuses less on famous people, and more on the ideas.
6. [Grayling 2016, p247].
7. [Grayling 2016, p253].
8. [Harari 2014, p284], [Gleick 2011, p7].
9. [Mumford 1934, pp44–45].
10. [Mumford 1934, p14].
11. [Landes 1969, p2].

12. [Harari 2014, pp60–62].

13. [Harari 2016, p98].

14. Bregman's quote "for the later progressives, curiosity actually became a virtue" is my translation from the Dutch "voor de latere vooruitgangsdenkers werd nieuwsgierigheid juist een van de grootste deugden" from his book *De Geschiedenis van de Vooruitgang* [2013, p143]. See also James Lovelock, who links "the original sin" to the discussion on our impact on the environment, and states "our fall happened *because of our knowledge*" [2019, p56].

15. As Stephen Fry insists, in *Mythos – the Greek myths retold*: a 'pithos', πίθος, was a "kind of glazed and sealed earthenware jar" in ancient Greek, which was misread for 'pyxis', πυξίς, which means box [Fry 2017, p133]. Though that seems a cylindrical vessel – anyway, jar it is then.

16. *The Printing Revolution in Early Modern Europe* is the abridged version of Elizabeth Eisenstein's earlier two volume work *The Printing Press as an Agent of Change* which was published in 1979. As Eisenstein states in the preface "my treatment is primarily (though not exclusively) concerned with the effects of printing on the written records and on the views of the already literate elites. Discussion centres on the shift from one kind of literate culture to another (rather than from an oral to a literate culture)" [1983, pxii].

17. Elizabeth Eisenstein discusses this plot in a later publication: *Divine Art, Infernal Machine – the reception of printing in the West from first impressions to the sense of an ending* [2011, p2].

18. [Mumford 1934, p26].

19. Rutger Bregman mentions Stigler's Law in his book *De Geschiedenis van de Vooruitgang* ('the history of progress') [2013, p66]. Bregman though, didn't seem to realise that the connection with Merton was intentional ("helaas bleek de wet van Stigler al vijfentwintig jaar eerder te zijn ontdekt door de socioloog Robert Merton"); Stigler in fact published the article in a book in honour of Merton, and on purpose named the law after himself – to prove the law [Stigler 1980].

20. [Gombrich 1995, pp281–285].

21. [Eisenstein 1983, p148].

22. Publications by Marshall McLuhan preceding *The Gutenberg Galaxy* [1962], such as the articles 'Culture Without Literacy' [1953] and 'The Effect of the Printed Book on Language in the 16th Century' [1957].

23. [McLuhan 1953, pp10–11], [McLuhan 1957, pp6–11].

24. [McLuhan 1957, p7].

25. Actually there are two images of J. S. Bach but they are almost identical, by the same painter, Elias Gottlob Hausmann, the first made in 1746 and the second two years later, which is the best conserved and therefore most reproduced image. This was two years before Bach's death in 1750, and it is often noted that the image gives a one-sided view of the character of Johann Sebastian Bach as a very serious-looking person. His music reveals many other sides of his character. See Eric Siblin's book *The Cello Suites J. S Bach, Pablo Casals, and the search for a baroque masterpiece* [2009, p11] (thanks to Jos Mulder who gave this book to me).

26. [Elsworth-Jones 2012], [Potter 2019].

27. In Sir Ernst Gombrich's *Little History of the World* [2005, p193], [2011, p220].

28. David Landes, *The Unbound Prometheus* [1969, p1, p41].

29. [Mumford 1934, p75].

30. Humphrey Jennings, *Pandæmonium* [1985, pxxxviii]. Jennings assembled his extensive notes in the decades before his untimely death in 1950, and the notes were posthumously published (with a foreword of his daughter) and introduced by his friend the sociologist professor Charles Madge, in 1985. *Pandæmonium* recently gathered prominence as "the book behind the Olympic opening ceremony" in London in 2012 (as a review in *The Guardian* put it). I am grateful for the suggestion for this book by Sir Tim Smit (17 March 2017 when visiting UTS). Tim Smit is the founder of the Eden Project [Smit, 2001], the very impressive greenhouse park site which I visited in 2004.

31. [Landes 1969, p41].

32. [Grayling 2016, p210].

33. [Heldring et al. 2015].

34. [Bregman 2013, p102].

35. [Gombrich 2005, p284], [2011, p244]. Most of the *Little History of the World* was actually written in the 1930s! Ernst Gombrich, as a young art historian in Vienna who had just completed his PhD, responded to a call of a publisher to write an educational text for young people about world history. Apparently, after sketching the outline and agreeing with the publisher, he set out to write one chapter a day – Sundays were for leisure, and for wooing Ilse Heller, as Leonie Gombrich recollects about her grandparents in the preface of the English edition. Gombrich finished the whole book in the agreed six weeks, and it was published in 1936 as *Eine kurze Weltgeschichte für junge Leser* ('a short history of the world, for young readers'). Gombrich and his family fled to England because of the Nazis, and became the famous and eminent art historian which we know from *The Story of Art* [1995] and *Art and Illusion* [2002] (inspiring this and other chapters). The book was translated into many languages and republished in German in 1985, but never in English until Sir Gombrich and friends finally got round to it and the wonderful *A Little History of the World* appeared in 2005 (posthumously), and in a hardcover illustrated version in 2011 – the page numbers refer to each.

   It is such a wonderful book. According to Leonie Jr. the original was banned by the Nazis not for racial reasons but because it was "too pacifist" [2005, pxvii], [2011, pxv]. And so it should be.

36. [Jennings 1985, p127]. Blake's poem inspired the song *We Work the Black Seam* by Sting on his first solo album *The Dream of the Blue Turtles* in 1985 (and in other versions, for instance on the live album *Bring on the Night* recorded later that year, and rearranged for symphony orchestra on *Symphonicities* in 2010). The song links the coal mining (the 'black seam') to the dangers of the waste products of nuclear power (though why he sings about Carbon-14, which is primarily used for dating artefacts in archaeology, and not particularly harmful, we don't know).

37. [Jennings, p48].

38. [Bregman 2013].

39. Frank Trentmann discusses the development of materialism in his book *The Empire of Things – how we became a world of consumers, from the fifteenth century to the twenty-first* [2016].

40. [Grayling 2016, p258].

41. [Mumford 1934, p88], [Landes 1969, p279].
42. [Landes 1969, p4], [Bregman 2013, p103].
43. [Blockley 2012, Ch4].
44. [Horowitz and Hill 1980, pp2–4]. All this is covered in any basic electronics textbook, in my case among others *The Art of Electronics* by Paul Horowitz and Winfried Hill, the first edition that we used in my electrical and computer engineering degree in the 1980s; there is a second edition of 1989 and a third edition of 2015. Other useful sources of basic electronics are *Make: Electronics – learning by discovery* by Charles Platt [2009], which is very accessible and hands-on, as reflected in the note on the front cover "a hands-on primer for the new electronics enthusiast" and further encouragement of "burn things out, mess things up, that's how you learn". Another favourite in the Maker movement is the *Getting Started with Arduino* manual about electronics and microcontroller programming (3rd edition), by Massimo Banzi and Michael Shiloh [2014]. See also the book *Physical Computing* by Dan O'Sullivan and Tom Igoe [2004].
45. [Nørretranders 1998, pp4–5].
46. James Gleick writes in *The Information* that in the 1940s computers "were, as ever, people" [2011, p208].
47. [Gleick 2011, p3].
48. A good overview of the use of the early electronic technologies to create musical instruments is in *Electric Sound – the past and promise of electronic music* by Joel Chadabe [1997], who passed away in May 2021. In my PhD thesis I describe some of these instruments too, with a focus on the interface [Bongers 2006, pp53–55].
49. [Nørretranders 1998, pp10–13], [Gleick 2011, pp269–272].
50. [Mumford, 1934, p89].
51. [Gleick 2011, pp13–27].
52. [Gleick., pp19–21].
53. [McLuhan 1962, 1964].
54. [Kelly 2016, p173].
55. [Yunkaporta in *Sand Talk - how indigenous thinking can save the world.* [2019, p15].
56. [Landes 1969, p277].
57. [Trentmann 2016].
58. *The Arts and Crafts Movement,* by Elizabeth Cumming and Wendy Kaplan [2002, p7].
59. *Bauhaus* by Frank Whitford [1984].
60. *Bauhaus – crucible of modernism* by Elaine Hochman [1997, p267].
61. [Markillie, 2012].
62. [Landes 1969, p4].
63. This development was already discussed for instance in an article, 'Implications on Design of Rapid Prototyping' [Hague et al. 2003].
64. [Oosterhuis 2003, 2011].
65. [Landes 1969, p2], [Diamond 1997, p259].
66. [Mumford 1934, p68].
67. [Mumford 1934., p84].

68. [Kayzer (ed.) 1993, pp166–181]. I vividly remember Freeman Dyson discussing this topic with Wim Kayzer on Dutch television in 1993. Kayzer made a range of evening-long TV interviews (!) for the VPRO station with several thinkers and scientists under the title *Een Schitterend Ongeluk* (Dutch translation of Stephen Jay Gould's expression related to human evolution as 'a glorious accident'). This title supports Kayzer's fascination with human evolution in many facets, as reflected in his choice of thinkers to interview (in addition to Dyson and Gould, Oliver Sacks, Daniel Dennett, Stephen Toulmin, and Rupert Sheldrake). The TV programmes have been broadcast in other countries including the US, and since 2004 have been available as a set of DVDs. A book with transcripts was published by the VPRO [Kayzer (ed.) 1993] (only available translated in Dutch, unfortunately, but it is a better translation than the original subtitles which were often very loose translations).

69. [Dyson 1979, p29].

70. [Dyson 1979., pp26–28].

71. [Kayzer (ed.) 1993, p179].

72. [McLuhan 1964, p155].

73. Rutger Bregman wrote the book in Dutch which was published in 2019 as *De Meeste Mensen Deugen – een nieuwe geschiedenis van de mens* ('The Majority of People are Decent – a new history of humans'), and published in 2020 in English as *Humankind – a hopeful history*. The translation, by Elizabeth Manton and Erica Moore, is very accurate though it has lost some of Bregman's quirky and jovial humour (which would only be appreciated by people from the Netherlands anyway – such as myself, I find it a joy to read his books). The text of the English version has also been slightly extended in some sections, and an index has been added which was missing in the original. My references to this book are given with page numbers referring to the original Dutch version [2019] and the English translation [2020].

74. *Utopia for Realists* [2017] is the title of the English translation of *Gratis Geld voor Iedereen – hoe utopische ideeën de wereld veranderen* [2020], originally published in 2014. The title of this third edition translates as: 'free money for everyone – how utopian ideas change the world', and makes a plea for the universal basic income.

75. [Bregman 2019, pp43–65], [Bregman 2020, pp21–40]. An excerpt from the book which covers the story of the Tongan boys was published in *The Guardian* on 9 May 2020, combined with an article by Jonathan Freedland, 'Rutger Bregman: the Dutch historian who rocked Davos and unearthed the real Lord of the Flies' (both appeared in *The Guardian Weekly* edition of 15 May 2020).

76. As the academic Diana Baumrind already indicated at the time, in the journal article 'Some thoughts on ethics of research - After reading Milgram's "behavioral study of obedience"' [1964].

77. The Milgram experiment is covered in *De Meeste Mensen Deugen* [Bregman 2019, pp203–223] and *Humankind* [Bregman 2020, pp160–178]. The conclusion that the behaviour was not about 'following orders' is later in the chapter [Bregman 2019, p216], [Bregman 2020, p172]. Gina Perry's book was first published in 2012, I used the 2013 revised edition. The link to the

Candid Camera TV show in the late 1950s and 1960s is on page 15, and mentioned in a transcript of a discussion between Milgram and one of the subjects [2013, p105].

78. [Bregman 2020, p170], or in Dutch "het kwaad … moet zich steevast vermommen als het goede" [2019, p213].
79. Interview with Charlie Brooker, by David Smith, published in *The Guardian* on 13 May 2020 (and in *The Guardian Weekly* on 22 May).
80. [Bregman 2019, p122], [Bregman 2020, p90], [Yunkaporta 2019].
81. The discussion is summarised by Tim Flannery in *The Future Eaters* [Flannery 1994, pp180–186], and also covered by Jared Diamond in *Guns, Germs and Steel* [1997, pp42–44] (but his is entirely based on Flannery's account). Yuval Noah Harari writes about it in *Sapiens* [2014, pp70–83].
82. [Cane 2013], [Attiwill and Wilson (eds.) 2003, pp30–32].
83. [Flannery 1994, p241].
84. [Flannery 2010, pp32–39], [Lovelock 2019].
85. [Lovelock 2019, p12]
86. Lewis Mumford *Technics and Civilisation* [1934, p51].
87. [Mumford 1934, p43].
88. [Mumford 1934., p51], emphasis added. Page 51 is my favourite page in the book.
89. [Nørretranders 1998, p308]
90. Mark Weiser wrote the influential article 'The Computer for the 21st Century' in *Scientific American* [1991].
91. [Harari 2016].
92. [Lovelock 2019, pp85–86].
93. See also Tor Nørretranders's discussion on the limitations of science, particularly on holism and reductionism [1998, pp355-360.
94. David Freedman in a magazine article 'Why Scientific Studies Are So Often Wrong: the street light effect', with a slightly different version of the joke [2010].
95. Lewis Mumford in *The Myth of the Machine* – technics and human development [1967].
96. [Crutzen 2002].
97. [Lovelock 2019, p37].
98. As discussed in a recent article in *The Guardian*, 'The Anthropocene Epoch – have we entered a new phase of planetary history?' [Davison 2019].
99. [Waters and Zalasiewicz et al. 2016].
100. [Bryson 2003, p193].
101. [Lovelock 2019, p38].
102. [Bryson 2019].
103. [Bryson 2003, p282].
104. [Cane 2013, p13].
105. [Cane 2013, p26].
106. According to an article in *The Guardian*, '"Glum future for the platypus": why the elusive mammal is disappearing under our noses' (by Graham Redfearn), referring to research by researchers from the University of Queensland published online on 7 January 2021 [Brunt et al. 2021]. The researchers conclude their article with the sobering statement "The platypus is a species that could disappear without us noticing."
107. [Harari 2014, p82].

Chapter 6

Communicating
Information, representation, and semiotics

Our mental tools shape the way we are as human beings and how we approach the world, as much as our technological tools do. These tools, such as language and arithmetic, scaffold and structure our representation and understanding of the world – both the environment as well as our inner thoughts. This of course also influences the way we represent and think about the world, through the reciprocal influence of language and other tools as discussed in Chapter 3 *Creating*. Representations are always interpretations, and their ability to structure comes with a certain regimenting and a risk of not fully grasping the complexity of what is represented. The influence of language and representations on our thinking is discussed in this chapter.

The chapter looks at how information is coded, and discusses the three main modes of information (re-)presentation: abstract (symbolic), mimetic (iconic), and the thing itself (the object or entity). This leads to the main modes of interaction, **symbolic** (such as written text, spoken language), **iconic** (mimicking that which is represented, often through images), and **manipulation** (with the actual object or physical presence).

Our perception of the outside world, as well as our inner thoughts, is shaped by languages. This is so natural to us that some thinkers suggest that language is innate, as mentioned in Chapter 2, and which will be discussed further in Chapter 10, *Thinking*.

Just as industrialisation had a profound influence on the way we make things, in the factory-based manufacture of tools, products, and technologies, there was also a development in information and knowledge – the way it was stored, communicated, disseminated. It developed from its physical forms (clay tablets, papyrus, parchment, paper) which were inherently mostly static, to dynamic forms using mechanical means (such as the printing press with movable type), electrical means (such as the telegraph), electronic (telephone, radio, tape recorders), to digital electronic and interactive computers. All these technologies and media made the world seemingly smaller, the "global village" as media theorist Marshall McLuhan famously put in the 1960s, and liberated the information sources – potentially at least, of course many media have been monopolised.

DOI: 10.1201/9781315373386-6

This chapter looks into information and communication, and particularly into **semiotics**, the study of signs, how information is coded and communicated, and to some extent also into **linguistics**, the study of language(s), and **semantics**, the study of meaning. Meaning is not just transferred, it can be further developed (intentionally or not) by the mutual influence of the participants involved, the sign systems used, and the tools and media that are applied. The potential for extracting and constructing meaning is everywhere in our environment; information is everywhere. Not only in the form of human-made information, which can be both intentional (designed) or accidental (interpreted) representations, generally our environment is full of information from many sources, that we are often not even aware of. We live in a dynamic environment, an ecosystem that is constantly changing by the forces of nature, animals, growth, technology, and human action. This is discussed in detail in Chapter 9 *Experiencing*.

PRESENTATION AND REPRESENTATION

It is essential to make a distinction between **presentation** (the object or entity itself), and **representation** – an image or other form of modality that *refers to* the object or entity. An object (or a process, or any entity) can **present** itself, communicating directly through its presence, including its **affordances** which have the potential to inform the environment (perceiver) of its possible functions, potential for actions, and particularly **manipulations**. It can also be **represented** in a coded form, as a **sign**. A sign 'stands for' something. It can be anything from a copy or image (a pictogram, **iconic representation**) which mimics (aspects of) the object or process, to an arbitrary sign (such as a word, **symbolic representation**), the meaning of which is established by a known convention between those involved in the communication or interaction. These are not necessarily separate categories; this chapter will discuss representation as a continuum or spectrum from mimetic (iconic) to abstract (symbolic). It is a spectrum, from realistic likeness to that what is represented, to fully abstract.

This chapter describes the **three main modes**, which are relevant for communication, knowledge (re)presentation, and interaction: **manipulative** (physical interaction), **mimetic** (iconic), and the **abstract** (symbolic).

These representational interaction modes are not only common in semiotics, but also linked to theories of human development. The development of mental activities of children takes place in several distinct stages, as described by the Swiss psychologist Jean Piaget in the 1930s, Lev Vygotsky in Russia around the same time, and Jerome Bruner in the US later in the 20th century.[1] The development of abstract thinking starts with a **kinetic**, sensorimotor, or practical intelligence stage (to use Piaget's terms) of physical manipulation of objects, also called the enactive phase (Bruner's term). The next stage of development is the **iconic** or mimicking stage, and finally the **symbolic** stage is reached, where abstract and learned signs are used. These theories formed the basis of the now widespread graphical user interface (GUI) through the research at Xerox PARC in the early 1970s, and still the main computer interaction paradigm. This is discussed in more detail below in this chapter.

COMMUNICATION – SIGNS AND MEANING

We can communicate through signs, a process which has been studied in great detail in the disciplines of linguistics, semiotics, semantics, literature, and communication studies. A sign can be visual, auditory, physical, olfactory (smells, scents), or anything else that can be perceived through our senses. Signs are organised in *codes*, manifested in for instance human languages, musical notation, number systems, and mathematics.

Signs are essential for human communication, they facilitate the processes of externalising our thoughts, our ideas, our mental world, and concepts. Symbolic language (using abstract signs) also allows for another level of reasoning. Analogous to the information communication model presented below, we can see that there is an internal state of the transmitter, externalised as a sign in a coding system, and received (and interpreted) by a receiver. The connections in this system mean that we can also use signs and language in our mental world, in addition to concepts and thoughts. Of course this meaning-making process is much more complicated than that, so we will come back to it several times in this chapter. It is also important to notice that, although signs are essential for the transmission of a message as stated before, meaning is often not just 'transmitted' but is a result of an ongoing process of exchange between sender and receiver, switching roles, developing ideas, and including existing knowledge, including unconscious knowledge which is often embedded in shared cultures.

There are about 6,000 to 7,000 different languages in the world at present, and a much greater number of dialects, which are variations within a language but are still mutually understandable. (Though when living in Maastricht in the south of the Netherlands I struggled to understand Mestreechts, the local dialect, and people told me that even between villages in the area there were different dialects which made communication difficult. And that was just within the province; crossing country and language borders, which do not necessarily coincide, within tens of kilometres one would have to speak Flemish (Vlaams, a dialect of Dutch), Wallonian (Waals, a dialect of French), or German).

The number of languages worldwide is rapidly declining, and particularly Indigenous and oral languages have suffered. In the last few hundred years many have become extinct, never to be spoken again, and some have only a few hundred or even a few dozens of speakers left. The densest language area is New Guinea, with about one million people and 1,000 languages – partially due to the geographical conditions.[2] The Australian continent was until about 10,000 years ago connected to New Guinea, which had about a similar sized population (estimated at 1 to 1.5 million people). Before the British colonisation started, Australia had about 250 different languages. And they are actual languages, not dialects, as linguist Bob Dixon explains in his recent book *Australia's Original Languages – an introduction*.[3]

Information and communication theory

In *The Mathematical Theory of Communication*, Claude Shannon and Warren Weaver in the late 1940s presented a communication model as a structure with an information *source*, which *transmits* a *signal* (containing the information in a *code*), through a *carrier* (or medium), to a *receiver* reaching the *destination*.[4] The signal in the carrier can pick up *noise*, in engineering terms, which distorts or adds information. This is why most communication systems have a level of redundancy built in. For instance when listening to a conversation in a noisy environment, or reading text from a damaged piece of paper, the information can potentially still be understood.

The model is a good starting point, also for technological perspectives, but interaction and communication are a lot more complicated when multiple participants, entities, roles, environments, channels, and feedback loops occur. As James Gleick points out in his book *The Information – a history, a theory, a flood* from 2011, the model that Claude Shannon developed was unique in that it attempted to define what information actually is, and quantify it (in bits, see below). As the expression goes, 'information is not knowledge', something which Gleick emphasises throughout his book and particularly towards the end, where he adds "and knowledge is not wisdom".[5] A hierarchy emerges: (raw) **data**, the actual codes or bits, **information**, the potentially meaningful data, **knowledge**, which can be applied, and has utility, possibly leading to **insight**, and aiming to reach **wisdom**, or the reflective insight through intelligence. To get to the wisdom level, often a lot of effort, time, practice, reflection, and experience is required. (Information is sometimes also called 'intelligence', or more colloquially, 'intel'; the 'I' in government security service names such as the CIA and MI5 stands for 'intelligence'. And to microchip manufacturer Intel we might suggest that 'processing is not intelligence'.)

As discussed in Chapters 3 and 4, in oral cultures the elders are keepers of the knowledge and wisdom. This is practised and applied, using physical objects and narrative structures to support the memory of knowledge by organising the information.[6] In literal cultures knowledge is stored in libraries, traditionally in paper sources such as books, later in analogue electronic storage, and recently in digital formats which has increased opportunities for interactive storage and retrieval (and in that sense actually closer to oral culture, as Marshall McLuhan often pointed out in the 1960s, anticipating the information age).

Shannon and Weaver in *The Mathematical Theory of Communication* distinguish three levels of problems in communication: technical, semantic, and effectiveness. In the publication, the chapter by Claude Shannon (a reprint of his earlier paper with the same title) was mostly concerned with the first level (the technical), while the two introduction sections by Warren Weaver discussed the second level (semantic) and third level (effect). The interesting aspect (and very radical at the time) of Shannon's model is that it is independent of the *meaning* of the messages. After acknowledging that "Frequently messages have *meaning*", as Gleick puts it, Shannon "then showed it to the door"[7] as follows: "The semantic aspects of communication are irrelevant to the engineering problem."[8] As Danish science writer Tor Nørretranders put it, Shannon was primarily interested in communication as "a theory for the transmission of information not its meaning."[9] In the context of Shannon's model, information is defined as the amount of choice in a message, the smallest amount being the bit (binary digit) – a choice between two states (yes/no, +/-, high/low, etc.). Shannon and Weaver acknowledge the mathematician J. W. Tukey, a colleague of Shannon at Bell Labs, who first suggested the word 'bit',[10] but the radical move of Shannon was to propose the 'bit' as the unit to measure information (data).[11] And although a bit can have only two states, which seems very limiting and potentially reducing everything to dichotomies, as mentioned in Chapter 5 (in the section on digital technology), if there are enough bits a very high precision can be reached (to store audio on the compact disc, the analogue signal is sampled in numbers of 16 bits, 2^{16} or 65,536 distinct values each, at a frequency of 44.1 kHz). Nowadays we are very used to this notion of 'data' as a quantity, for instance with one's mobile phone plan comes a certain amount of bandwidth over time, for instance 2 GB (gigabit) of data, or the expression 'I have used all my data'.

Looking more closely at the nature of 'information', Shannon's idea was that if a step in a process limits the choice of the next step (for instance, in many Western languages including English the letter 'q' will always be followed by the letter 'u') it is said that the information density is less. From this follows that random letters have a higher level of information than structured language, which is in contrast to the lay definition of information. In information theory there is a clear distinction between information and meaning. Information is defined as the ratio between what is known and what is not known. The Dutch media writer Arjen Mulder, in his book about media theory,[12] gives a good example from nature: a leaf with an irregular edge contains more information than a leaf with a clear edge, because from one chosen point on the irregular edge it is difficult to predict where the next point on the curve will be, while the clear edge is more predictable. "Information is surprise", as Gleick puts it.[13] Shannon introduced the notion of Information Entropy as a measure for uncertainty in a message, similar to the notion of entropy in physics as the degree of randomness or disorder, as discussed in Chapter 5. A formula was presented to establish the 'channel capacity' as a measure of the amount of information (in bits per second), depending on the bandwidth of the system, the signal level, and noise level.

Though the model has its roots in engineering, and in cryptology (Shannon worked on military code breaking in the Second World War, just like Alan Turing in Great Britain),[14] the model was intended to describe human communication, direct or mediated; after all, Shannon worked for the Bell phone company.

Around the same time, in the late 1940s, the field of cybernetics emerged as proposed by Norbert Wiener, studying and developing models of control systems.[15] In a control system when the signal is returned to the transmitter it can act as a *feedback loop*, a term that was already used for electronic systems that connected the output of a circuit back to its input. Feedback is important as it allows for adjustment, tuning, and improvement, and is as such studied in engineering as well as in biology. Feedback can be *negative*, to regulate the circuit or system, bring it under control (the 'governor' mechanism that made the steam engines actually usable was an early example of this principle), or *positive*, which reinforces the signal and makes it go out of control, such as when the sound output of a speaker is picked up by the connected input microphone (or a guitar – electric guitar players in the 1960s enthusiastically experimented with this technique, particularly Jimi Hendrix was a master of playing with feedback). Positive and negative are mere technical terms, not implying any value judgement (as illustrated with the deliberate choice of examples above). Different types of feedback play an important role in interaction models, as discussed in Chapter 9 *Experiencing* and Chapter 12 *Understanding*.

Curiously, however, in Wiener's cybernetics, information is regarded as order, the opposite of 'Shannon information' which is based on entropy, or disorder. This is also our everyday notion of information, that it is about organisation.[16] But that 'information' is about meaningful information, or knowledge, one level up in the hierarchy presented at the beginning of this section.

Meaning making

The Shannon and Weaver communication model works well for the transfer of information, however in order to understand interaction we need to look at how meaning is made. They do acknowledge the importance of the context: "one of the most significant but difficult aspects of meaning".[17] Later, in the field of semiotics, context (pragmatics) became one of the key areas of study, which is discussed below. This can also be seen in the work of anthropologist E. T. Hall, who studied human communication in a range of different cultures, and realised that some cultures relied more on context than others. A high-context culture relies more on the stored (and shared) knowledge to establish the meaning, whereas a low-context culture relies more on the transmitted information to establish the meaning. This was first published in 1976 in his book *Beyond Culture*, and further developed in the 1980s in *The Dance of Life – the other dimension of time*.[18]

For the purpose of understanding interaction, this notion of context and dynamic processes of meaning making is important. In many cases, a message is not just generated and transmitted adding a potential for influence of noise that influences the message, but when received it is also interpreted. The phrase 'beauty lies in the eye of the beholder' refers to this. The interpretation depends on the personal characteristics of the receiver, the context in which the transmission or interaction takes place, and the cultural background, to name a few. Different branches of semiotics, semantics, and communication theory have different emphases and approaches to this complex topic, but the main point in the context of understanding interaction, as stated before, is that *meaning is made as part of the process*. It is entirely possible that the *meaning emerges* out of the process of interaction, unforeseen and unintentionally, but appropriate(d).

Meaning in this view, therefore, is a dynamic entity, it can change over time. Music is a good example of this, everyone will hear a piece of music differently depending on their musical knowledge, taste, culture, mood, etc. and each time when one listens to a piece of music new things can be discovered (particularly with complex, composed or some improvised music). Re-reading a book after several years shows how one extracts different meanings out of the same material, due to the increased knowledge and experience in the world of the reader (one hopes).

It is worth noting that it is potentially confusing that in the English language the meaning of the word 'meaning' can be different whether it is a noun (as discussed in this section) and a verb; 'to mean' has to do with someone's intentions.[19] In most other languages separate words are used, for instance in Dutch 'betekenis' (meaning as a noun) and 'bedoeling' (meaning as a verb).

Signs and semiotics

In this section the study of signs, **semiotics**, is discussed. This includes a historical overview of how ideas and insights came about, and the key figures in these developments.

Semiosis is the process of transferring meaning from or about an **object**, via a **sign**, to a receiver. Signs are used to convey or establish meaning; **semantics** is the field of study concerned with meaning.[20]

A distinction in representational signs that has been made is between **icons**, which mimic the object, and **symbols**, which are abstract and arbitrary, established by convention. A further category is the **index** or symptom, signs which are caused by the object or process. For instance, smoke

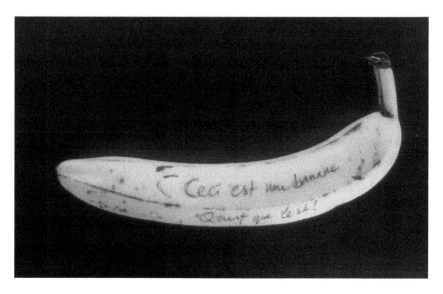

Figure 6.1 This is not a banana (2017).

is a symptom of fire, and therefore an index. (Although perhaps it can be more useful to see fire and smoke as one event, one entity.) The notion of the index is important in linguistics, as it often occurs that a sign points at something (or at another sign), creating a chain of meaning through referral, discussed in sections below in this chapter.

These terms can be illustrated with an example. Take for instance a banana, the actual piece of fruit (as shown in Figure 6.1). The banana is the **object**. A picture of a banana **represents** the banana; it is not the object itself. This is what the Belgian surrealist painter René Magritte famously showed in the late 1920s with his painting of a pipe, with a statement written below of *"Ceci n'est pas une pipe"*.* Indeed, this is not a pipe one is looking at, it

---

* Magritte created a range of paintings of different objects, such as an apple and a pear (no banana though), under the title *La Trahison des Images* (1928–1929), often translated in English as *The Treachery of Images* and indeed it is about how the image betrays or deceives the viewer by pretending it is the real thing. The 'pipe' painting has been often copied and paraphrased in popular culture and other domains, for instance, the cover of Oliver Sacks' book *The Man Who Mistook his Wife for a Hat* [1985] has a picture of a hat with the subtext "Ceci est ma femme" (illustration by Paul Slater). And even on the website of La Fondation Magritte, where it reads: "Ceci est un site web". Another example of Magritte is the painting *Le Palais des Rideaux III (The Palace of Curtains 3)* (1928), in which we see two paintings (in identical polygon shape frames), one depicting the sky, and the other one with the word 'ciel' in it. The fact that it is a *painting* of two paintings, adds another layer of representation. It is a favourite example of Ernst Gombrich to illustrate the difference between a figurative (iconic) representation and an abstract (symbolic) representation (in the preface of the 5th edition of *Art and Illusion* [Gombrich 2002, pxviii], and on the cover of the 'Millennium Edition').

is a picture of an object (the pipe), and therefore a mimetic (iconic) representation.* The word 'banana', spoken or written, is entirely abstract and arbitrary, it doesn't sound or looks like a banana yet it is used as a label, a symbolic representation. This symbolic (abstract) representation is different depending on the language, 'banaan' in Dutch is similar and so is 'banane' in German and French, but it is 'platana' in Spanish, and 'pisang' in Bahasa Indonesia, but all refer to the same piece of fruit (more or less).

The smell of the banana might be considered an index – but I prefer to see it as a property of the actual object, in its multimodal manifestation (I can see it due to the light that reflects of it, I can smell it, it can be touched). And if you took a banana, as instructed at the beginning of the paragraph, you can now eat it. If you didn't, and were just imagining one, this could be called the phenomenological banana – this is discussed further in the Chapter 9 *Experiencing* and Chapter 10 *Thinking.*

History and overview of semiotics

As mentioned above, semiotics as the study of signs is closely related and overlaps with the field of linguistics, the study of language(s). Not everything studied within the discipline of semiotics is relevant for understanding interaction. In order to apply and extend some of the semiotic studies in the field of research and design of the interaction between people and technology, it is particularly suitable to explore certain subfields of semiotics. Examples are: the developments of insights in **social semiotics**, bio- and **zoosemiotics** (how animals communicate in often radically different ways compared to humans), **multimodal semiotics** (how different representational modes and sensory modalities are used), **sonic semiotics** (using sounds for communication, including music, and soundscapes as an environment), and **ecological semiotics** (which links the ecological approach to perception, discussed at the end of this chapter, and in depth in Chapter 9 *Experiencing*).

In the next sections some of the history and basic ideas of semiotics are presented and discussed.

C. S. Peirce – philosophy and logic

Although the study of signs as the carriers of meaning is probably as old as human language itself, the first person to structurally analyse sign systems, and develop frameworks that are still used in the field of semiotics, was C. S. Peirce (pronounced as 'purse') towards the end of the 19th century in the US. Charles Sanders Peirce was a philosopher and developed many ideas on logic, but didn't publish everything he wrote during his lifetime, so a lot of what we know of his impressive breadth and depth of thoughts is extracted from his notebooks, articles, and correspondence, a process that

---

* *Ceci est une note de bas de page.*

is still going on. He used the term *semeiotic*, referring to the work of the philosopher John Locke in the late 17th century.

Peirce's basic proposal was a triadic theory of signs, approaching the sign as a whole consisting of three parts: the **representamen** (the sign itself), its relation to the **object** (an actual object or a mental concept), and the **interpretant**. The latter is somewhat abstract, an interpretant can be the interpreter (the receiver of the information) but is more widely defined as a 'potential for interpretation', what it means in the mind of the perceiver.[21] This is an important notion, that will keep coming back in the discussions on semantics: through the interpretation, the receiver of the message is part of the process of meaning making. Peirce further proposed that an interpretant in itself can become a sign (a representamen), potentially leading to an extended spiralling chain of meaning making (or 'unlimited semiosis' as he called it).

The basic signs that are used in communication as defined by Peirce in 1867 and as mentioned before, are the **icon**, which mimics the object or concept it refers to, the **symbol**, which has a more or less arbitrary relationship with the object or concept (assuming shared knowledge about this convention), and the **index** or symptom, a sign which is caused by the object or process (it points to, or reveals, the object, as it were). As mentioned in Chapter 1, the GUI paradigm in HCI reflects these categories, including the indexical, which are representations to do with system processes (indicating progress in tasks such as file copying, establishing a network connection, CPU use, etc.).

Peirce further distinguished between the **immediate object** (as represented by the sign) and the **dynamic object** (which effectively produces the sign, and is independent of it).

Depending on the structure of the coding systems (for instance, symbolic language), signs could be organised in **types** (for instance, particular words in the language) and their instances that occur as **tokens** (words appearing on a page).

Acknowledging that communication between entities doesn't only take place in an explicit, coded, and highly structured way (although this is what is mostly studied in later semiotics), Peirce identified levels of communication: **firstness**, the kind of direct, unmediated and uncoded interaction and mimetic communication, **secondness**, which uses indexical signs and is about the relationship between the objects, and **thirdness**, the purely symbolic and which is mediated (and encompassing the other levels). Firstness is similar to the direct perception (particularly the notion of affordances) in the ecological approach of J. J. Gibson[22] which is discussed at the end of this chapter, and in detail in Chapter 9 *Experiencing*.

De Saussure – linguistic structures

Another founder of the field was the Swiss linguist Ferdinand de Saussure, at the beginning of the 20th century. De Saussure actually didn't publish much on the topic of what he called '*semiology*', but notes gathered from his many lectures were later transcribed and posthumously published as *Course de Linguistique Générale* in 1916 (translated into English in 1959 as *Course in General Linguistics*).[23]

The basic terms that De Saussure used are **signified** (the concept or meaning) and the **signifier** (the entity which carries the meaning, such as a word on a page, or a drawing), a dyadic model rather than the triadic model of Peirce.

Although De Saussure's contributions were very insightful and influential, he was also quite explicit about what he did *not* study in human language. De Saussure distinguished between what could be called the 'designed' language ('langue' in French) which is structured, well ordered, and intentional, and can be studied, and on the other hand the actual spoken language ('parole' in French, 'spreektaal' in Dutch) which is often messy, ever changing, and breaks rules, which he deemed too hard to study (it is also too hard for spell checkers on the computer or smartphone, they often 'correct' colloquial and creative use of language). This ruling out of some of the complexity as too hard to study is of course typical in Western science, we seem to study what we can study (however complex) and tend to ignore some really difficult parts, see the discussion on the 'streetlight effect' towards the end of Chapter 4, but at least De Saussure was upfront about it. For the same reason of narrowing down the scope to make it manageable, he focused on the language system at a particular point in time (synchronic), rather than taking into account how a language changes over time (diachronic). He also excluded how context, existing knowledge, culture, etc. influenced the meaning of signs (pragmatics).

An insightful discussion of De Saussure's ideas is presented by Robert Hodge and Gunther Kress, who proposed including precisely those areas that were omitted (identified by going through "Saussure's Rubbish Bin", the metaphor they use) in the field of study of social semiotics.[24] This is why the social semiotic approach has so much potential for application in understanding interaction.

Further semiotic developments early in the 20th century

These 'founding fathers' had many followers for various reasons, resulting in the first half of the 20th century in somewhat separate streams of semiotics in the US, with Peircean admirers such as Charles Morris and later Thomas Sebeok, and De Saussure's semiology in Europe with people like Jean Baudrillard and Roland Barthes.

The analogy with developmental psychology later in the 20th century is perhaps a coincidence, with key figures from Switzerland (Jean Piaget), east coast US (Jerome Bruner), and Russia (Lev Vygotsky), with individual roles; respectively: insightful/flair/dogmatic, wide ranging/deep/applicable, and original/recalcitrant/oppressed). The Soviet counter in the semiotics case came from Voloshinov and colleagues from the 'school of Bakhtin', advocating an emphasis on social interaction and studying the actual spoken language (the 'parole' that De Saussure had dismissed).[25] However, due to their Marxist leanings they were being suppressed in their own country (by Stalin) as well as ignored by the West in the 1920s. Voloshinov's ideas eventually became popular in Western semiotics in the 1960s.

Ogden and Richards – Literature

Early in the 20th century two Cambridge academics, C. K. Ogden and I. A. Richards, studied the relationship between signs, content, and meaning from the perspective of English literature, in the book *The Meaning of Meaning – a study of the influence on language upon thought and of the science of symbolism* which was first published in 1923.[26]

Firmly realising that meaning was not a fixed thing that would be transmitted in communication to a receiver, but that meaning is established in the process, they took Peirce's work as a starting point.[27] The 'semantic triangle' they proposed, still often reproduced in semiotic textbooks, place the thing or concept represented in the bottom right corner (the 'referent'), the representation (the 'symbol') in the bottom left, and the resulting meaning at the top ('thought or reference'). These are similar to the terms Peirce used, the object, representamen, and interpretant. In the Ogden and Richards triangle the axes are used to indicate the relationships and possible interpretations and connotations between these elements.

From the background of literature, and particularly poetry, Ogden and Richards emphasised that words often have multiple meanings, not only all the ones listed in the dictionary but also the shifting meaning as a result of the context, or pragmatics (as anyone who tries to learn a new language will be confronted with, in particular). Marshall McLuhan, who studied with Richards at Cambridge University in the 1930s, was deeply impressed by this notion and realised the importance of it and expanded it in later media.[28] While this was relevant for meaning making in literature as a prominent medium at the time, for instance when the reader of the poem constructs a unique version of the content in their mind, McLuhan realised this would be even more so with the then new and still emerging electronic media and technologies.

Karl Bühler – Communication

What is interesting about the older explorations in theory development to describe communication, is that the concepts presented are usually only known to us in technical or media terms, while in those days the technologies were not yet invented, or under active development. Examples are the frameworks and theories that Shannon and Weaver, and Norbert Wiener were exploring and developing in the US, as discussed above. In Europe, an interesting communication model was developed in 1934 by Karl Bühler, a

German Gestalt psychologist specialising in speech and education, particularly child development.[29] His Organon model is useful because it emphasises the different intentions in producing a sign, which can be a symptom, a signal, or a symbol. A spontaneous utterance, which is more a reflection of one's internal state ('Ausdruck' in German, or expression), is a symptom. When the sign is an attempt to direct a mostly one-directional communication, such as an order or a request ('Appell') it is a signal. This is the part of the communication that the later model of Shannon and Weaver focus on, however, with Bühler's model there is an ecological aspect to the notion of the 'Appell', in that the receiver has to be susceptible to it. The word 'appeal' in English has a similar meaning, with the purpose of driving an action in the receiver. The third part of the model is about representation; when the sign refers to an object or concept, in the situation where one is communicating with the intention to explain or represent ('Darstellung') it is a symbol. The latter can be a sign anywhere on the continuum from recognisable depiction (iconic) to fully arbitrary (symbolic).

Charles Morris – areas of semiotic study

In the 1930s Charles Morris in the US developed Peirce's ideas further, and he remained an influential figure in semiotics in the 20th century. Morris proposed to categorise the field of semiotic (as it was then called in the US) into the "dimensions of semiosis": *semantic, syntactic*, and *pragmatic*. The semantic dimension would study the relation between the sign and the object (or denotation, this about the meaning), the syntactic dimension is about the relationship between the signs (the syntax or order, how signs relate to each other), and the pragmatic dimension which deals with the interpretation (how the process of meaning making is further influenced by the receiver, or how signs relate to a person).[30] Morris's approach was strongly influenced by behaviourism in psychology at the time, with its focus on 'stimulus-response' studies. Although useful and well developed, it has limitations in its application in a more ecological approach which is needed in design for interaction.

Morris himself wrote in his 1946 book *Signs, Language, and Behavior* "These terms have already taken on an ambiguity which threatens to cloud rather than illuminate the problems of this field."[31] The three identified areas of semiotic study (and they really are areas of study, rather than the proposed "dimensions"): semantics, syntactics, and pragmatics, were in later literature interpreted as categories of semiotics, or worse, modes of signifying, or even sign categories. This confusion still exists in some popular accounts of semiotics. However, as areas of study, these categories (Morris's "dimensions") can still be relevant.

Robert Hodge and Gunther Kress wrote that Peirce's observations on semiotics "are potentially far more subtle and fluid than the mechanistic theories of those who claim to follow him".[32] They specifically mention

Morris, which makes sense, but they seem to also imply Thomas Sebeok. This is understandable, some of the work by Sebeok (who studied with Morris) is a bit 'mechanistic', but his work has a much broader scope as discussed in the next section. This is why some of Sebeok's work can be extended very well into the frameworks for designing technology for interaction.

Thomas Sebeok – establishing the discipline

The second half of the 20th century brought the field together, particularly through the efforts of Thomas Sebeok (who, as were many in this field, was from an eastern European background and had moved (fled) to the US before the Second World War – such as Roman Jakobson).

The term *semiotics* became established from when it was first introduced by the anthropologist Margaret Mead in 1962, as the field which "in time will include the study of all patterned communications in all modalities".[33] This was during a discussion session on linguistics as part of a conference on Paralinguistics and Kinesics at Indiana University in May 1962, which brought together key speakers from cultural anthropology, education, linguistics, and psychology. This discussion session was following a paper on emotive language. It is interesting to note the context, which clearly pushed the boundaries of the field beyond what is common in linguistics (addressing emotion, movement, expression, culture, etc. was not commonly attended to in the linguistics of that time). As we can read in the (edited) transcript of the discussion, in the proceedings published two years later, Mead emphasises several times the need for a "word for patterned communications in all modalities", and suggests 'semiotics', which after a short discussion on using just the word 'communication' ("a word that is now muddled by the mass media"), seems to have become established.[34] In Mead's presentation, 'Vicissitudes of the Study of Total Communication', the term is further applied,[35] and it is reflected in the title of the book with transactions of the conference, *Approaches to Semiotics*. The theories and frameworks developed by Thomas Sebeok are certainly of relevance for design.

Sebeok is also known for his work on *zoosemiotics*, and wrote several articles about his studies of how animals communicate. Looking at how some animals communicate is a rich source for inspiration for design, for instance, the use of chemicals, electricity, ultrasound, light, and movement, in activities such as perception, expression, defence, or attack (or a combination of these activities). A selection of Sebeok's articles on zoosemiotics since 1960 was bundled in a book published in 1972.[36]

This work has led to a framework of *channels*, which are related to our sensory modes and media used, reflecting the insights and inspiration drawn from Sebeok's studies of animal communication and behaviour. The framework was originally described in an article on animal communication, and is often replicated as a diagram in semiotic literature.[37] Sebeok's framework first makes a distinction between communicating through *matter* and *energy*. Matter is categorised in *solids*, *liquids*, and *gasses*. Energy is categorised in *chemical* (further divided in *proximal* and *distal*) and *physical* (further divided in the main categories of *optical*, *tactile*, *acoustic*, *electric*, *thermal*, etc.). In the optical channel an essential distinction is made between *reflected*, *daylight*, and *bioluminescence*, and the acoustic channel is divided in *air*, *water*, and *solids*. This is a useful taxonomy, which can easily be extended and applied to interactive systems.

Elsewhere, in an article[38] that is concerned with linking Shannon and Weaver's communication model and other more engineering approaches with semiotics, Sebeok proposes to divide the channels into *bands*, as is common in engineering (frequency bands such as short wave and long wave in radio, UHF and VHF in TV, and different modulations techniques such as AM and FM – these are technical distinctions, however, and mostly have to do with how the technology is constructed to support different frequencies (inverse of wavelength) and modulation techniques). What Sebeok attempted to identify are actually channels, or modalities, and as such the proposal is useful: "the vocal-auditory band" (speech addressed to a listener), "the gestural-visual band" (meaningful movements, to be seen), "other modalities such as touch, smell, taste, and temperature", and "finally, what might be called manipulational-situational band" (handling mediating objects and spatial relationships). These 'bands' are actually quite appropriate, as a starting point, and examples of these modalities ('bands') can be roughly seen in HCI in speech-based interaction, gesture-based interaction, affective computing, and tangible interaction, respectively. However, a distinction between direct and mediated communication needs to be made, which we will return to in Chapter 12.

Sebeok then carried on with developing a combination of Bühler's model (see above) with Shannon and Weaver's in a double triangle of relationships, and the six functions of communication as defined by Roman Jakobson[39]: referential, emotive, phatic, conative, poetic, and metalingual (extended from Bühler's emotive, conative, and referential). Jakobson looks at relationship between an 'addresser' (source of a message) and an 'addressee'. He defines *emotive* as the 'expressive' function, more reflecting the state of the addresser than directed communication. *Conative* is about the explicit addressing of the receiver of the information. *Poetic* is about the message as such, and its (verbal) structure. *Metalingual* is communication about the communication, for instance, when asking 'do you know what I mean?'. The relevance of the context of a communication (and its contribution to the meaning-making process) is called *referential*. The *phatic* function is about social communication, rather than being about meaning, communications such as small talk, gossip, and 'chit-chat'. The main function of phatic communication is to connect, to establish a start or potential for further conversation, and discussed in more detail later in this chapter.

Informed by the insights gathered from studying animal communication, Sebeok made a distinction between 'language' and 'speech'.[40] In Sebeok's definition, 'speech' is symbolic and abstract communication, and he states that only modern humans (*Homo sapiens*) possess this ability. Although research has been carried out that shows the ability of some animals to learn symbols, Sebeok's work on zoosemiotics at the time (1960s–1970s) insists that this is limited – that there is no evidence for grammar, or the generation of new knowledge through the use of symbols. Sebeok also states that early hominins didn't have 'speech' (use of symbols), only language. However, as discussed in Chapter 3, Steven Mithen in *The Singing Neanderthals* [2005], presents a more insightful account on the development of human language and speech, convincingly demonstrating the possibility of speech (and gesture) and musical co-development: his acronym 'hmmmmm' encompasses holistic, multimodal, manipulative, musical, and mimetic to describe the evolution of human language (and speech, in Sebeok's terms). One chapter in Mithen's book (Ch. 8) looks at the more recent state of research in symbolic communication by primates, and he comes to the same conclusion as Sebeok, but his theory about early humans' communication is different. Also, recent insights from the field of ethology (the study of animal behaviour), shows that at least to some extent certain animals are capable of forms of symbolic communication including grammar, as presented for instance through numerous examples in Frans de Waal's books *Are We Smart Enough to Know How Smart Animals Are* [2016] and *Mama's Last Hug* [2019]. De Waal refers to this historical underappreciation of the complexity of animal behaviour as "anthropodenialism". Sebeok's work on zoosemiotics has to be seen in the context of attitudes and values of his time, and as such can still have relevance.

Sebeok's distinction between language and speech, and a further distinction of mediated communication (using writing, film, etc.) is worked out in his three 'modelling systems'. A modelling system is an internal 'model' of reality that a person (or animal) can have. In HCI it is common to refer to a 'mental model' that someone can have of (the workings and potential of) a system, and a well-designed interface supports the person in developing this mental model through which predictions can be made about the behaviour of the system. In modelling systems theory Sebeok proposes that each system is species specific, emphasising that species would each have their own modelling capacity. This is an approach based on the notion of the *Umwelt* in the work of the German biologist Jakob von Uexküll.[41] In Uexküll's book from 1909 *Umwelt und Innenwelt der Tiere*, about the relationship between the environment and the inner world of animals, Uexküll's defines the *Umwelt* as the environment in relation to person or animal, as an ecological concept, referring to what the subjective experience of the environment is to a specific species. The notion Umwelt is similar to the definition of the 'niche' in ecology, but broader and more meaningful.[42]

Sebeok's *primary modelling system* concerns the communication using 'language', in the sense that it is about communicating directly and through mimicking rather than symbolically. The *secondary modelling system* is about symbolic communication, such as speech (in Sebeok's definition). The *tertiary modelling system* is about writing and more complex structures, concerning human culture. It involves media, such as books, paintings, or television programmes. Participants in the interaction use models about their environment and the characteristics of the system in order to communicate. This distinction between direct interaction (primary/secondary) and mediated interaction (tertiary) is useful, and will be worked out further in Understanding Interaction Frameworks for Design in Chapter 12. Some authors would label all language a medium,[43] but in the context of understanding interaction it is more useful to distinguish between *direct*, using our own 'devices', our own organs and *mediated*, using media, which augment or translate our utterances and expression, and require an interface.

The modelling systems (primary, secondary, and tertiary respectively) roughly correspond to Peirce's *firstness*, *secondness*, and *thirdness*.[44] Sebeok's labelling of primary and secondary, language and speech, is partially about the distinction between mimetic and symbolic communication. Speech is similar to Saussure's *langue* and what is studied in the structuralist approach.[45] There is essentially a hierarchy in the modelling systems, not so much in a sense of importance or relevance, but in that each system encompasses the lower systems. So the symbolic language ('speech' in Sebeok's term, the primary system) encompasses mimetic and direct communication (the secondary system), and the tertiary system as a culturally relevant aggregate encompasses the primary and secondary systems. The systems could also be defined in broader terms, the first as *interaction*, the second as *communication*, and the third as *media*, which will be applied in Chapter 12.

THE ROLE OF CONTEXT IN MEANING MAKING

As stated before, in many cases meaning is made in the very act of communicating. The concept ('object' in Peirce's terminology) can be influenced by the way it is represented ('representamen'), by the interpretation ('interpretant'), and by the context. Umberto Eco even calls it the *referential fallacy*, the idea that the meaning is only determined by the object referred to.[46] Eco, who was a professor at the University of Bologna wrote extensively on semiotics, language, mediaeval history, and philosophy, particularly his first major work in English, *A Theory of Semiotics* published in 1976. He also wrote several historic novels such as *The Name of the Rose* from 1980 as mentioned in Chapter 5.

Meaning is not only conveyed through the signs or codes themselves, a message can be interpreted using the receiver's knowledge, cultural values, conventions, etc., and this process takes into account the context, references, and circumstances. As mentioned above, anthropologist E. T. Hall identified cultures on a continuum from high to low context.[47]

In summary, meaning making is to an extent personal, contextual, dynamic, situational, and iterative. There is so much more information involved, mostly implicit and unconscious, than just the information explicitly exchanged in spoken language and writing – this is shown in the situation of a participant anywhere on the continuum of ASD (autism spectrum disorder) might not grasp the implicit elements of the communication. And of course many interactions with computer systems fail because of the lack of awareness of context in the design for interaction. This is further discussed in Chapter 12.

Pragmatics

The study of this extracoding (meaning outside of the codes) is part of what is called *pragmatics*.[48] However, *pragmatism* is also an approach in (particularly American) philosophy that emphasises the practical application of knowledge and insights, first formulated by C. S. Peirce in the late 19th century and further established by the psychologist William James and philosopher John Dewey.

In semiotics, pragmatics relates to the role that the interpreter plays in the processes of meaning making. It is one of the three key areas of semiotics as identified by Charles Morris (semantics, syntactics, and pragmatics) as mentioned above.

To give an example of the role of context or pragmatics: I remember from a presentation by some of my linguist colleagues at IPO (the Institute for Perception Research, a collaboration between Philips and the TU Eindhoven, in 1997) the phrase "the man hit the boy with a stick",[49] which requires outside knowledge (extracoding) to resolve the inherent ambiguity of who is holding the stick, and therefore whether the boy deserved to be hit!

Danish science writer Tor Nørretranders in *The User Illusion* introduced the term *exformation*, reasoning from the strict definition of Shannon of information, as all the discarded information in a communication[50]. This is a really insightful concept, it is about the effort a participant in a conversation puts into reducing the information exchanged in the communication, by anticipating and understanding the other participant's knowledge and context. While in semiotic literature extracoding and pragmatics are discussed as something essential yet potentially somewhat confusing and mysterious, Nørretranders emphasises that it is a completely normal and inevitable aspect of communication, yet that it takes a fair amount of effort by the participants to establish this. This is the power of the right expression, often found in poetry where a sparse amount of words conjures up just the right image in the receiver's mind, in all its intended complexity. It is what Marshall McLuhan often refers to as the *all-at-once*.

Strata – four domains of practice

A useful concept to discuss the influence of context (or practice) is the use of *strata*, as defined by Gunther Kress and Theo van Leeuwen in their 2001 book *Multimodal Discourse – the modes and media of contemporary communication*. Each stratum defines a stage in the process from sender(s) to receiver(s), from articulation to interpretation. Strata or stages can overlap, but in the example below a more sequential case is described. In the book, Kress and Van Leeuwen first make a useful distinction between *content* and *expression* (which is the manifestation, in this approach). The content can be separated ('stratified') in *discourse* and *design*. Discourse is the context in which the communication takes place, including social and cultural aspects such as attitudes, ideas, values, evaluations, and purposes. It is therefore strongly related to the shared or group knowledge (as discussed in Chapter 10), both in implicit and explicit modes. In their stratum of *design*, the term refers not so much to the 'shaping' meaning of the word, but rather the 'planning/composing/devising' meaning. The *expression* layer or medium is further subdivided into *production* and *distribution*. Production refers to the material form the expression takes, usually the carrier. Distribution is about how the content gets disseminated, transmitted, carried. Combinations of all strata are possible, for instance in improvised jazz or avant-garde improv music the strata of design (composition) and production (performance) are merged. It is also important to note that the strata do not imply an order or hierarchy.

Kress and Van Leeuwen use the term *'medium'* in the way it is common in describing the media of an artwork (e.g. oil on canvas, bronze, mixed media) and the tools used. It is the material form and instruments through which the expression manifests itself, and often with new media the only physical form is the *carrier* (the disk, the electromagnetic field, the light in an optical fibre). Medium in the context of *Understanding Interaction* includes the non-physical aspects, it is the medium that facilitates communication.

The purpose of the notion of strata is that these layers can have their own potential for signification. This is something that in a different way was already identified by Marshall McLuhan with his aphorism "the medium is the message", indicating that a medium can influence and even *transform* the content of the sender. McLuhan looked at advertising in the 1950s to develop this notion.[51] Indeed, with the introduction of mass media, new media, and hypermedia there are many examples of this notion, discussed in more detail below in this chapter.

INDEXICAL SIGNS: POINTERS, SYMPTOMS, AND TRACES

As stated above, the main categories of representational signs (as defined by C. S. Peirce) are *mimetic* (iconic), *abstract* (symbolic), and *indexical* (symptoms, and other pointers referring to what is represented). This range of indexical signs, which point at other signs or entities, is discussed in the next sections.

Index

This third form of sign commonly mentioned, the index, is a sign that is caused by or related to the referent. This can be a *symptom*, meaning it is related to an object of process (for instance, yawning is a symptom of tiredness, or of boredom), or even caused by the object or process (for instance, the causal relationship between fire and smoke). In the textbook *Communication – an introduction*,[52] the term *signal* is used as a synonym for *index* which is potentially confusing because of the way the term is used in signal theory (units of transmission, rather than signs).[53] A signal is usually defined as a sign with only one meaning (not to be influenced by interpretation), for instance, a road sign (at least that is the intention ...), or an electrical signal controlling a machine.

The word index is used not so much in the sense of organising, but in its original meaning, which is that of pointing (hence we have an index finger). Returning to the earlier example of the banana, the presence of fruit flies may indicate (point to) the presence of a banana (and probably the state in which it is – the less fresh, the more chance for fruit flies to be present) (after all, as the linguists' joke goes, "fruit flies like bananas" – this has two possible meanings depending on the interpretation).

Deixis

The category of signs of indices, is particularly relevant as the type of sign that points to something else. In this case multiple signs are working together to convey meaning, often using multiple expressive channels simultaneously (usually speech and gesture). In linguistics these are called *deictic* signs, such as a word that is dependent on its context for its meaning, through referring to other words. Umberto Eco, in his discussion on indices in *A Theory of Semiotics*, almost entirely focuses on this deictic nature (rather than symptoms as discussed above) and uses the term *verbal shifters*, the meaning of the (verbal) sign *shifts* due to a (necessary) additional action.[54] The common example is the utterance of 'this' or 'that' while pointing at something, proximal or distal respectively, in a combination of a verbal and kinesic action. (An indexical sign can also be *anaphoric*, which is similar to *deictic* but it refers to something within the same utterance, for instance 'I like bananas so I am going to eat one.')

One of the earliest examples of multimodal HCI was a project in the late 1970s, called '*put that there*', combining a verbal expression with gestural expression.[55]

Traces

Objects and processes leave *traces*, which can be seen as indices displaced in time, remnants of past events, which are important for studies from archaeology to forensic science. These traces can be very useful as cues of people's intentions, their wishes and needs, in the remnants of past behaviours. In design this is often used, for instance around 1970 the architect Jan Gehl proposed the use of traces as a driver for urban planning for public spaces.[56].

In an ongoing project, *Traces – reading the environment*, I have studied various forms of urban and interior traces related to past use, and potential for reading these expressions of people's, including tracks created in an unplanned way – sometimes called 'desire lines' or in Dutch 'olifantenpaadjes' (elephant paths). These tracks are often shortcuts, a bottom-up approach to finding efficiency, often more effective than the top-down approach of planners. There is also the cultural aspect, of a strong component of civil disobedience in these expressions, and can be seen as the physical outcome of the accumulated choreography of individuals.[57] This practice also strongly suggests that the designed environments should ideally always support a dynamic and ever-changing range of uses. This principle can be applied to technology design in general, there are many illustrative examples of people adapting and appropriating technologies, and potentially it could be applied in software design as this is the most flexible technology. Although this potential was already identified in the early 1990s,[58] many current interfaces are still lacking the support for people to customise and adapt.

The *Traces – reading the environment* project includes lectures, a series of interactive artworks, and a short publication.[59] My fascination with urban traces in Australia led to a range of pieces (photographs, video, sensors, and sound) presented in a small exhibition in 2010, *Traces*, see Figure 6.2.

This included patterns created through video manipulations of the elaborate skid marks of car and motorcycle tires, which were interpreted in musical phrases (using feedback guitar, electronics, and extended techniques), *Skid Scream*. Other elements were *Pavement Painting* videos on the floor and photos on the wall, based on found traces of 'builder graffiti' which is spray-painted on the pavement and road surfaces to reveal the hidden infrastructure (electricity, water, gas, etc.), which often results in remarkably colourful patterns. And in contrast to much of regular graffiti, 'builder graffiti' seems to be actually legal.

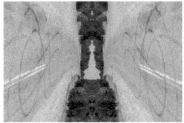

Figure 6.2 Images from the Traces exhibition (DAB Lab Gallery, Sydney, 2010): Skid Scream (left) and Pavement Painting (right).

Often the indices or traces are properties of the object or process itself, for instance, the smell of the banana is an example of direct perception of the actual piece of fruit, in the olfactory rather than the visual sensory modality. So, although these traces are caused by the object, we can say that they are actually part of the object. The smell of the banana is a property of it, the way it manifests itself, an essential part of being a banana. Objects can manifest themselves as well by making sound (usually as a result of a process, like cracking a nut, a ticking clock, a banana that gets squashed). Another example would be rain – not only can it be perceived by looking at the actual raindrops, but also by the sound the raindrops make when falling on a tin roof, the circles of the drops falling in puddles (the latter are traces of earlier rain), or it can be inferred from observing the moving of wind screen wipers on passing cars or pedestrians carrying umbrellas.

SEMIOTIC RESOURCES AND FRAMING

To refine the concepts related to modes of representation, Gunther Kress and Theo van Leeuwen describe the sub-modes as *semiotic resources*, or *parameters*.[60] These include for instance the paralinguistic elements and other subtleties and varieties of expression. In Van Leeuwen's chapter 'Parametric Systems – the case of voice quality' in *The Routledge Handbook of Multimodal Analysis*[61] he discusses the many aspects of the human voice (pitch, stress, timbre, loudness) which are all important for expression and meaning making. These voice quality aspects were earlier identified as 'dimensions', having semiotic potential in the book *Speech, Music, Sound*,[62] and are now presented as a parametric system. The semiotic potential of 'loudness' for instance, can be easily demonstrated by uttering the same phrase by whispering, or by shouting (and everything in between). Similarly, using different pitches for the same utterance will demonstrate the role of pitch in meaning making.

In his PhD thesis *Making Things Louder - amplified music and multimodality* of 2013 sonic scholar Jos Mulder (supervised by Van Leeuwen) presents a solid argument for the semiotic potential of electronic amplification. Where most people may think of amplification as just making sound *louder*, Mulder (based on his extensive practical knowledge as a live sound engineer) demonstrates how the experience (and therefore potential for meaning making) of the sound changes when amplified.[63] Amplification is considered a semiotic resource.

This is also an excellent example of what Marshall McLuhan in the 1960s meant with his famous aphorism "the medium is the message"[64]: the medium has a potential for influencing and even transforming the meaning of the message. He also talked about 'hot' and 'cool' media, terms that amalgamated some of the key factors in media, such as participation and bandwidth – "a hot medium is one that extends one single sense in 'high definition'". A movie in the theatre is 'hot', while a cartoon is 'cool', and "speech is a cool medium of low definition, because so little is given and so much has to be filled in by the listener."[65]

McLuhan illustrated this convincingly with his later book *The Medium is the Massage*, a collaboration with graphic designer Quentin Fiore in 1967. In the publication, the semiotic resources of graphic design (such as typography and lay out, and the images) are used as an integrated 'message' with the text.[66]

Other examples can be found in the visual arts, any painter or video artist (or curator) would know how much the *framing* of the image matters. The meaning of the work can change depending on its presentation, this is why attempts have been made in the art world to unify and standardise the presentation (white walls and certain lighting conditions in the gallery, dimensions and framing in the cinema, concert halls, performance theatres, etc., with supporting roles of curators, projectionists, sound engineers, etc.). My *Videowalker* explorations since the early 2000s (from solitary *Dérives* to public performances) are intentional investigations into the role of the *framing* by re-situating the video projections in real time, dynamically interacting with the environments by projecting video content in response or dialogue with its surroundings.[67] This project is about liberating the projector, which is usually bolted to the ceiling, and carry it around finding more interesting and appropriate projection surfaces than the rectangular frame of the fixed screen.

Art has a long tradition of reframing, of 'making strange' what seems familiar, in order to create new perceptions and insights. One example is (Australian artist) Ron Mueck's sculpture *Dead Dad* (created in 1996/7), which while being a life-like (a slightly inappropriate term in this case) representation of the artist's then recently deceased father, due to its small scale creates an alienating effect. The image is extremely intimate, but the framing creates a distance which perhaps makes it bearable. When I first experienced it, in the *Sensation* exhibition of Charles Saatchi's Young British Artists collection in 1997 (see below in the section on Value Creation and Authenticity in Contemporary Art), it elicited an experience of rapidly zooming out, as if I was ascending. This effect is less strong in pictures showing the work in context, and even entirely absent in the pictures of the work in isolation.[68]

Design thinker Kees Dorst prominently advocates the role of (re-)framing in design processes, as a powerful approach to problem understanding (rather than a premature focus on problem solving). His recent book on *Frame Creation – create new thinking by design* [2015] presents a theoretical grounding, and applications in a range of design fields, from product design to urban design and social design. Reframing is a fruitful and even essential strategy to go past the boundaries of the 'frame of mind' and fixed mindsets when approaching a problem space.

McLuhan wrote about the medium and the message in the early 1960s, when mass media gained prominence and he realised that increasingly we see the world through media. It is a mediated reality, not the real reality. We all have had opportunities to compare the two (or more) realities, for instance the beach in the brochure and the actual beach when visited, the view from the bedroom of a house for sale first seen on the housing website, a film adaption of a favourite book (remediating), meeting social media friends IRL (in real life). McLuhan would be delighted to know he didn't need to be around for the emergence of "reality TV", where participants become celebrities just by participating in the medium, becoming the message as it were. Some people might even take the mediated world as the real world, or at least as the preferred world, and base their opinions and behaviours on this.*

MIMETIC REPRESENTATION

As mentioned above, an important category of signs is the mimetic or iconic representation. This iconic representation can have a varying level of fidelity with regards to the object or entity mimetically represented, from a photograph, painting, drawing, sketch, or ideogram. An icon of an object can also be an object, for instance, a three-dimensional sculpture or a 3D print based on a computer model, it can be a meaningful object in itself, and/or it can be a representation that refers to something else. This is discussed further below, but here it is good to look at this notion of fidelity. At first one would be tempted to consider a photograph of an object to be the most realistic representation (in 2D at least) of that object. However often a painting can potentially be *more* realistic, as the painter can incorporate ('code') more knowledge and interpretations into the painting, as discussed for instance by art critic John Berger in *Ways of Seeing* in the early 1970s.[69] As a medium, of course the role of the painting as a

---

* Not only the warning of many science fiction stories and movies such as *The Matrix* and *eXistenZ*, but most clearly shown in *The Truman Show*, the 1998 film directed by Peter Weir and written (and co-produced) by Andrew Niccol, in which Jim Carey plays the eponymous character that is unknowingly trapped in a reality TV show watched by millions.

faithful or as an interpretation of reality has always been changing. Since the introduction of the medium of photography (the technology was first introduced early in the 19th century, and improved over the next decades), painting was increasingly liberated from its obligation to faithfully represent. Painting eventually developed fully abstract modes, in addition to the figurative. Along the way, several styles were explored, particularly from the early 20th century.[70] Impressionism sought to explore interpretations of perceived realities, Expressionism included the unperceivable expressions, Cubism showed different perspectives simultaneously and combined multiple viewpoints spatially, Dadaism had elements of anti-art, Surrealism deliberately distorted reality and included the imaginary and unconscious, Pop Art reflected and reframed everyday objects often in an exaggerated manner, and Conceptual Art represented ideas rather than objects, often in activities and happenings. Meanwhile photography and film developed as media for artistic expression on their own.

Onomatopoeia – sound words

In the sonic modality, iconic (mimicking) sounds are called *onomatopoeia* or "mimic words" as the linguist Barbara Lasserre describes them in the book *Words That Go "Ping" – the ridiculously wonderful world of onomatopoeia* [2018, p4]. In contemporary English there are not many words whose sounds refer to the object they represent, though my son pointed out that a wok makes a wok-like sound when you bash it (thank you, Rufus) (from the name in Cantonese, 'wohk', it is different in Mandarin, and in Bahasa Indonesia it is 'wadjang', a slightly different pan, but they all sound the same).

Many examples of onomatopoeia can be found in the names of animals (or machines), somewhat resembling the sounds that they make, for example, the word 'duck' (in English, though it is 'eend' in Dutch, 'canard' in French, and 'pato' in Spanish). It is a mild fascination in linguistic circles that the sounds animals are supposed to be making seem different in different languages. The rooster in particular, whose call I heard only too often when I was little, is supposed to say 'kukeleku' in Dutch, 'cock-a-doodle-doo' in English and 'coco-rico' in French, while it actually doesn't sound like any of these.* The cow sound words are usually a bit more accurate,

---

* David Sedaris in his 2004 book *Dress Your Family in Corduroy and Denim* comments on 'coco rico' "which sounds like one of those horrible premixed cocktails with a pirate on the label", but when he tells people what the rooster sounds like in his language they "look at me in disbelief". Sedaris writes how he uses the question of "What do your roosters say?" as a conversation starter, a good example of phatic communication. When travelling in the US he asks about gun laws, but as "firearms aren't really an issue in Europe" here he asks about "barnyard animals". Another question he uses is "When do you open your Christmas presents?", and this phatic opening turns into a meaningful exchange – as he shows by writing most of the chapter on the response to this question in the context of the Netherlands where there are indeed some truly remarkable and certainly questionable (to outsiders, not to us of course) aspects of our particular version of the folklore; one of these aspects is reflected in the title of the chapter 'Six to Eight Black Men'.

'moo' (English) or 'meu' (French) or similar in many languages, except for Dutch cows – they go 'boe' (pronounced 'boo' in English).*[71]

Of course there are also many examples of iconic (and symbolic) representations in other sensory modalities, such as smell (perfumes, food) (as indices as discussed above, but also as codes[72] and as proximal cues,[73] and touch (shapes, braille alphabet which is discussed in the section on Tactual Perception in Chapter 8 on sensory modalities).

Metaphoric representation

A common way to explain something that is unknown is by comparing it or referring to something that is well known, as a metaphor. Metaphors work by association and analogy, or parallels, and are a powerful way of describing a complex concept with one aptly chosen reference, mapping from one domain to another.

In poetry, which often works through association to conjure up images and concepts, metaphors are a powerful technique of conveying thoughts more immediately. If well chosen, metaphors have the potential of linking thoughts directly, as it were. It can be described as "the summit of poetic imagination", with which a poet can carry "a concept away from its natural environment into an entirely different realm" as linguist Guy Deutscher describes it in *The Unfolding of Language*.[†]"Metaphor' means 'taking something from one place to another', or more literally to 'carry across' in Greek and as Deutscher points out by showing his photo of a removal van in Athens with the word ΜΕΤΑΦΟΡΕΣ on its side. He comments that this is not "because they are advertising courses in creative writing".[74]*

---

* Ever since coconut husks were used to imitate the sounds of horse hooves (a common Foley artist technique, apparently) in Monty Python's 1975 movie *The Holy Grail*, everyone who has seen the movie (or at least a fragment of it) will have this association of coconut sounds signifying horses. (The budget for the film was £229,575, which obviously didn't allow for real horses.) The knights of the round table in the movie pretend to ride an imaginary horse, supported by the coconut sounds and other noises provided by their servants (Patsy, King Arthur's "trusty servant", is a cameo role of co-director Terry Gilliam throughout the whole movie. Gilliam also plays "the old man from scene 24" who recurs several times, and himself as animator feigning a heart attack to get out of a difficult plot, a typical deus ex machina twist).

† A great explanation of metaphors is in the 1994 movie *Il Postino* (co-directed by Michael Radford) in which the modest and shy postman (played by co-director Massimo Troisi) wins over the girl of his dreams by wooing her with *metafore*. He learned about how to use metaphors from the Chilian poet Pablo Neruda (played by French actor Philippe Noiret), who is in exile on the Italian island where he delivers the mail. Guy Deutscher uses this example too, although he refers to the book *Burning Patience* by Antonio Skármeta that the film was based on – with the same dialogue [2005, p115–116].

Metaphors in language

In *The Unfolding of Language* Deutscher shows how prominent metaphors are in everyday language. He devotes a whole chapter to it, 'A Reef of Dead Metaphors',[75] in itself a metaphor which ties in with his wider use of geoscientific terms to describe dynamic processes of language development, such as erosion. Language constantly changes and evolves, and it always has done that (otherwise it would never have been developed in the first place, as Deutscher intended to show with the subtitle of his book, *the evolution of mankind's greatest invention*). This makes it hard to study of course, as mentioned earlier De Saussure had explicitly suggested to exclude the diachronic (changing) and that only the synchronic (language as fixed in one moment in time) could be studied systematically.[76] But the dynamic is relevant for understanding the evolution, the unfolding, and the potential for future directions. Deutscher shows how much of our everyday language is in fact metaphorical, and that there is a constant development from the concrete (the real-world physical entities the metaphors refer to) to the abstract. This is similar to the development of icons (mimetic) to symbolic (abstract) in semiotics. In the book *Metaphors We Live By*, George Lakoff and Mark Johnson [1980] already showed how much of language depends on metaphors, and how it influences meaning, often in ways that we are not aware of.

The point which Deutscher makes, that language develops and improves (rather than degrades) through evolution and change, is also the topic of linguist David Shariatmadari, who writes "linguistic decline is the cultural equivalent of the boy who cried wolf, except that the wolf never turns up." As Shariatmadari points out, people have complained in all ages about the so-called decline in language use, but it is in fact the necessary development that enables its expressive capabilities to keep up with the demands of contemporary life and culture.[77]

Metaphors in the GUI

In the original GUI (Graphical User Interface) concept metaphors played a strong role, with the 'desktop' as a work space, the filing cabinet image as a storage space, and the 'trash' as a place to delete files (on my current system it is called 'bin' which sounds less trashy but it is actually a bit confusing). As discussed in Chapter 1, this way of interacting with computers was first developed at Xerox PARC (The Star) in the early 1970s, based on the ideas of Douglas Engelbart among others in the 1960s, and later popularised by Apple (Lisa, Mac) in the early 1980s. One of the first books to reflect on the design for interaction and interfaces for computers was *The Art of User Interface Design*, edited by Brenda Laurel and conceived by Joy Mountford (then manager of the Human Interface Group at Apple) published in 1990 as mentioned in Chapter 1. Tom Erickson reflects on the early use of metaphors in the GUI in the chapter 'Working with Interface Metaphors', stressing the importance of choosing the right (most appropriate) metaphor.[78] But he also points out that the limitations of the thing or concept referenced might potentially limit the abstract concept. In the same book, a chapter by Ted Nelson emphasises this limitation of using metaphors, after their initial helpfulness to get you from one domain to another, "the metaphor becomes a dead weight".[79] To take a common example, a real garbage bin will eventually be full, but the virtual bin ('trash') on the computer GUI might take much more items. The metaphor also leaves ambiguity about when it will be 'emptied'. Microsoft Windows' bin was presented as a recycling system, while of course the items don't get recycled but the space they used (bits on the hard disk storage) might be freed up for new purposes. These are all examples of the inherent problem with using metaphors, that the limitations of the concept can be misappropriated in the concept described. An even stronger example is the use of the 'trash' icon to eject a disk (or another connected computer) in the original versions of the Apple Macintosh GUI, which was their implementation of the WIMP (Windows Icons Menus Pointer) paradigm as developed by Xerox PARC. This was often confusing for people, you would feel tricked into fearing that you would erase your floppy disk (this was in the early 1990s) by putting it in the garbage bin. In the recent (Mac OSX) versions of the Apple GUI, the 'trash' icon quickly metamorphoses into a eject symbol when you start dragging a disk (it is a bit of a hack, this 'sleight of icon', but ok).

Another example is the 'folder', the icon image of a actual folder – this limits an object or a file to be only in one place, just as in the case of the real-world folder – Alan Kay, one of the inventors of the original WIMP paradigm at Xerox PARC, reflects on this limitation and suggested 'bins' (not the British macOS version of a 'trash' can, see above), that are "active retrievers that are constantly trying to capture icon instances that are relevant to them".[80] This can be done manually in later versions of the operating system, using pointers (which refer to a file, there can be many pointers); in Unix this is called an 'alias'.

Actually I often wondered at the time why there are not more metaphors and icons in the GUI referring to *actions*, through mimetic gestures for instance. There are a few of those now on the touchscreen and touch pads (particularly the multi-touch ones) such as swipe sideways for delete, pinch to zoom, etc. In the older GUIs, most metaphors refer to objects and things, even if they 'do' things, like an icon of a 'printer device' prints, the 'trash' deletes, and a picture of a 'floppy disk' saves. Which is a good example of icons becoming symbolic rather than mimetic, even I haven't actually used floppy disks for decades – and I read an anecdote of a teacher showing a floppy disk in class, where a student thought it was a 3D printed version of the icon! (Ironically, my former UTS colleague and now professor at UniSA in Adelaide Ian Gwilt made a range of interesting artworks based on physical representations of icons, such as *Folderculture* around 2006 – he would have appreciated this!)

There are also many examples of metaphoric shapes in products, such as those designed by Philippe Starck for instance: with the shape of my Alessi cheese container *Mister Meumeu* from 1992 Starck refers to the head of a cow, with my Heller toilet brush *Excalibur* from 1993 Starck refers to a sword, my *Zeo* TV for Thomson from 1993 is overall very playful, with an antenna that looks like it can wag like a dog's tail, and of course Starck's *Juicy Salif* lemon squeezer for Alessi in 1990 (and yes, it *does* work as such, it is not just a statement or "conversation starter" as Starck himself described it), which shape is inspired by a squid but also refers to a 1960s idea of a space ship.[81]

Another example is the work of Naoto Fukasawa, who designed many products for Muji, and used the image of a shower fan for the shape of his CD player for Muji in 1999 – including the action metaphor of pulling the string to start and stop it.[82]

Skeuomorphs

An extreme form of metaphoric design is the use of *skeuomorphs*, where the shape of a contemporary object or even a building mimics an older form (or elements of it) as a relic, as ornaments. Examples of skeuomorphs are a mobile phone handset looking like an antique phone handset, or a contemporary building posing like an old one (even just through ornaments, 'classical trimmings'*). Early iPhone and iPad apps were a bit like that, with spiral bound note pads, and paper diaries, and there is music generation software that mimics old analogue synthesisers.[83]

A CONTINUUM FROM MIMETIC TO ARBITRARY – THE REPRESENTATION SPECTRUM

As stated above, an *icon* can have a varying level of similarity to what it refers to (a banana, for instance), from photograph, realistic painting, drawing, sketch, and finally reduced to a depiction of the most basic form, an ideogram that contains just enough information to be recognised as a (reference to a) banana. This is the sort of thing we see in traffic signs (particular in continental Europe with its many different languages; I have noticed that English-speaking countries tend to use text-based traffic signs

---

* As depicted in the 1949 movie *The Fountainhead* (based on the 1943 novel by Ayn Rand, who also wrote the screenplay), where we see how the clients of a young modernist architect (played by Gary Cooper) keep insisting on adding 'classical trimmings' to his sleek and ornament-free building designs.

more), in stations and airports, etc. Many 'icons' in our everyday environment, such as the heart which stands for love, or indeed traffic signs, are loaded with conventions and are not actually present in the sign itself. They are therefore more symbolic (abstract) than iconic (mimicking). It is therefore common in contemporary semiotics to treat the iconicity of representational signs as a continuum, from full likeliness (iconic, mimicking) to entirely arbitrary (symbolic, abstract).[84]

This Representation Spectrum is a very useful approach for design for interaction. Rather than seeing them as distinct categories, representational signs are seen as being on a *continuum*, from a realistic image (literal representation) at one end, and at the other extreme end actually abstract images that can only be understood with certain knowledge, and which depend on the cultural and social context. Categories are mere markers on this continuum. Potentially ambiguous icons actually have to be learned in order for them to be recognised, and in that sense the category of icons crosses over into symbols. The same is true for symbols, their meaning can be entirely abstract and arbitrary or there may actually be a similarity or relationship between the symbol and the meaning (particularly in spoken language). Another way to put it is that a sign is said to be *motivated* by what it refers to (a term De Saussure used). This is acknowledged by Robert Hodge and Gunther Kress in their book on Social Semiotics, who proposed to define this continuum as ranging from *transparent* (preferring this term over 'motivated', although Kress in his own books uses this term), if there is a clear relationship between the sign and the signified, to *opaque* in the case it is arbitrary.[85] This is, as stated above, contextual, and Hodge and Kress also emphasise that this is subjective in the sense that it can be transparent to the producer of the sign, but not to the receiver.

Umberto Eco was also critical about the classical tri-partition of signs, stating: "symbols, icons, indices: an untenable trichotomy".[86] He offers a number of useful additions and nuances, eventually leading to a new typology of four dimensions. I will try to focus on some key insights that are useful for design and understanding interaction. Eco shows that, as noted before, the categories of signs are better approached as continua (such as the later proposal of Hodge and Kress of transparent–opaque) and that often correlations and conventions are heavily dependent on culture and context. Eco refers to Morris who wrote about iconicity defining it as "similar in some respects to what it denotes" and that "Iconicity is thus a matter of degree".[87] In this part of *A Theory of Semiotics*, Eco gives an elaborate overview and in-depth discussion of what 'similarity' can mean, including replicability, doubles, replicas, sharing properties, similitude, convention, and function. Similitude, for example, can mean that an object is similar but in a different scale. Eco states that similitude is learned (and often produced with that ability of the interpreter in mind).[88]

A voodoo-doll could be seen as an early 'interactive' (if imaginary) (one hopes) example, where the doll represents a person and can be harmed in the person's stead.

Eco emphasises that in fact most icons have at least an element of arbitrariness. For instance, an iconic sign can be related to an object by similarity, by sharing properties, by being motivated, or by being analogous. The latter has to do with *fidelity of reproduction*, and Eco contrasts 'analogue' with digital (which is not meant in the technical sense, a representation can be analogous to what is represented, not in discrete approximation – as discussed earlier with the current state of digital technology, at least in some cases actually a higher fidelity can be achieved through digital rather than analogue technological means). Eco's discussions often mention that similarity can be based on a *pertinent feature*, and in a later book relates that to the discussion of *affordances* if the pertinent feature relates to the (potential) function of the object,[89] discussed in more detail below.

Another nuance Eco discusses is the notion of *doubles*. When we have two identical objects (which in itself is an interpretation as no two objects are ever exactly the same – but we define them as identical, see the discussion above about generalising), for instance in the case of two dice, two apples, or (one of Eco's examples) two Fiats 124 cars, one is not an icon of the other but doubles.[90]

This is related to the use of the terms *type* and *token* in linguistics, where the type is a generic term or the abstract descriptive notion, and the token is the actual instantiation of the type. For instance the notion of the word 'the', as a type, can occur on a page as a token. On this very page the token 'the' occurs many times. The ratio between the type and token, or the number of similar tokens, is a measure of variety in the language used. This notion of lexical richness could be used to analyse a typical speech use of the word 'like', like every other word, 'like, are you like, for real? Like for real for real?'. Lexical richness can be temporarily diminished in particular circumstances, for instance when a speaker is tired or absent minded, replacing certain nouns with the generic word 'thing'.

An extreme version is the language of the Smurfs (in English, or smurfen in Dutch, originally schtroumpf in French in the cartoons of Belgian author Peyo since the late 1950s), where any noun, verb, or adverb can be replaced with the homonym 'smurf' (in the appropriate variations and conjugations) – without the text losing its meaning. It makes use of the redundancy in human language. As such it is a "parasitic language", as Umberto Eco points out, it relies on the reader's knowledge of the typical structures and phrases in the original language.[91] Like, smurf, you know!

Eco relates the continuum from exact replica to arbitrary, to the *type/token ratio*, where the token is the icon and the type what it refers to (the *type* is abstract, it is a category, while the *token* is the concrete, the instantiation of the type, the sign). This is useful for design, with this necessary nuance of the notion of the icon, linking all the various strategies for iconic reference such as similarity, sharing properties, motivation, and analogy in one continuum. It forms one of four dimensions of Eco's proposed Typology of Modes of Sign Production.[92] In one of Eco's later books, *Kant and the Platypus – Essays on Language and Cognition* [1999], he develops this further and uses the term *cognitive type*.

HCI AND SEMIOTICS

Semiotics is not a common topic in HCI (Human–Computer Interaction) textbooks and literature. There is no mention in the HCI book by Alan Dix, Janet Finlay, Gregory Abowd, and Russell Beale nor in the HCI Handbook edited by Julie Jacko.[93] The Interaction Design book by Helen Sharp, Yvonne Rogers, and Jenny Preece only mention the work on 'Semiotic Engineering'.[94] There is a succinct introduction on semiotics in Orit Shaer and Eva Honecker's excellent overview of Tangible Interaction.[95]

Linguist Clarisse de Souza's 'semiotic engineering', developed from the early 1990s and presented in several publications in the early 2000s, applies the principles of semiotics on graphical user interface design. It is about the computer as a medium, which potentially facilitates (meta-)communication, but also very much related to computer science, which is also found in other research.[96]

The work of Carlos Scolari applies semiotic approaches to websites, blogs, and other new media with a focus on the user interface, and is also critically reflecting on the semiotic engineering approach.[97]

Since the discipline of semiotics studies in depth the communication between people (and their environment), it has some essential insights and tools to offer to understanding interaction, and its future potential for design.

Semiotic modes reflected in HCI

As mentioned at the beginning of this chapter, the developmental stages in human learning, as reflected in the semiotic modes, influenced HCI. These human learning stages were identified by Jean Piaget, Lev Vygotsky, and later Jerome Bruner. The terms they used are (Piaget/Bruner): symbolic/abstract, iconic/mimetic, and kinetic/enactive. Bruner's research identifies *mentalities* with each of these stages. The stages reflect the main modes of communication and interaction in general as described above, of manipulative, iconic, and symbolic. As stated before, the Graphical User Interface, developed at Xerox PARC (Palo Alto Research Center) in the early 1970s, popularised by Apple in the early 1980s (the Lisa and the Macintosh) and later Microsoft Windows in the 1990s, is still a prominent paradigm in design for interaction. The paradigm incorporates these modes of manipulative, iconic, and symbolic; the developments at Xerox PARC were inspired by the work of Bruner in particular, using the term 'enactive' for manipulative. In the decades of development since the early GUI, some further paradigms have been established reflecting the many human capabilities in communicating and manipulating and driven by technological progress, leading to increased effectivity and better experiences. Alan Kay's research team at Xerox PARC used this developmental theory (particularly Bruner's work as it was more concerned with mathematical language and information theory) to invent the GUI as we now know it: it has 'direct manipulation' with the mouse (enactive), icons (pictograms mimicking objects), and symbols (text in menus).[98] The mouse as a pointing device with two degrees-of-freedom movement was invented by Douglas Engelbart in 1964, which the Xerox team adopted and developed further.

Ironically, as Bill Verplank (design for interaction pioneer, see Chapter 1) often points out, the HCI field seemed to have developed in reverse: first interaction through symbols (command line interface), then interacting predominantly through visual iconic representations, and only recently the physical interaction (enactive) is more fully explored in the field of Tangible Interaction.[99]

This is in fact true for the whole development of computer science and AI (artificial intelligence), starting with symbols, later images, and finally physical robots. The effect of having a 'body', the physical presence of a machine, on its processing abilities is potentially very strong. In the further chapters the effect of having a body for human beings is discussed in detail, starting with the senses, perception, thinking (cognition, conscious and unconscious processes), and expressing.

Modes and modality in semiotics and HCI

In semiotics and linguistics, all the different ways of coding, the usage of different signs, are called *modes*. The common situation where multiple modes of communication are used simultaneously, then, is called *multimodal* (modal means 'of the mode'). This is the topic of several books.[100] In this definition, a page in a book containing text and various kinds of images is called multimodal.

In the psychology of perception, and (therefore) in HCI, it is common to use the term *modality* is used to refer to a sensory channel, as it were, such as the visual modality and the auditory modality, or an expressive channel, such as speech or gestures. In HCI since the 1990s the term *multimodal* refers to the situation where multiple sensory or expressive modalities are used, for instance, a combination of vision, hearing, and touching when manipulating an object or navigating a space.[101]

However, in semiotics and linguistics the term *modality* is used in relation to the (perceived or agreed) validity or truth of a message.[102] Detailed discussions on modality (in this definition in the semiotic discipline, which is related to truth) in visual modes can be found in *Reading Images* by Gunther Kress and Theo van Leeuwen, using the term 'modality markers', and for the auditory modes in Theo van Leeuwen's *Speech, Music, Sound* book.[103] This use of the term modality originated in Peirce's theories about logic. Peirce defined three main cases of modality, 'actually true', 'necessarily true' (as in inferred by logic), and 'possibly true' (as in a hypothesis). Robert Hodge and Gunther Kress add there are many more possibilities than these three cases,[104] and Umberto Eco points out that the continuum from icon to symbol (from transparent to opaque, see above) can also be seen as a scale of potential for lying. Indeed, the more abstract and arbitrary the sign, the easier it becomes to bend the truth. As Eco has famously put it, "Semiotics is in principle the discipline studying everything that can be used in order to lie."[105]

Depending on the language, 'modality' in the linguistic sense (as related to how true or false a statement might be, or the probability) can be expressed in *modal verbs*, examples in English (and similarly in German and Dutch) are: can, might, could, must, should, etc.

(Note that in Margaret Mead's original definition of semiotics as presented above, she uses the term modalities in this sense (sensory and expressive modalities), and so does Sebeok in his writings.)

In this book, therefore, I will always use the terms *sensory modality* (related to human perception) and *expressive modality* (related to human action), as an interaction 'channel', to distinguish it from the use of the term 'modality' in most semiotic literature, and use the term *multimodal* for the situation where multiple sensory and/or expressive modalities are used. This overlaps with the use of *multimodal* in semiotic literature, meaning multiple modes. In semiotic literature a *mode* is a *semiotic resource*, that is, a potential for signification, any way in which meaning can be communicated, in any sensory or expressive channel. The use of the term multimodal in semiotic literature is not specific about which *sensory modality* (or multiple modalities) are involved in the processes, nor which expressive modalities are used, and it pays little attention to other modes of being and communicating (implicit, involuntary, unconscious). We will return to this in Chapter 12, where frameworks for interaction are discussed, including modes of interaction, particularly the important notion of implicit interaction (as opposed to explicit), intentional or accidental presentation, peripheral (as opposed to focal) interaction, etc.

VALUE AND AUTHENTICITY

The issue of truth is worked out in detail by Umberto Eco in his book *A Theory of Semiotics*, in the context of replicability, illustrated with the example of money in the form of coins (which represent a value but also have a value of their own, due to the precious metals used) and banknotes (which have less value of their own).[106] It is even possible that due to inflation a coin becomes worth more than the monetary value it represents – the British 'sovereign' gold coin* is designed to have an intrinsic, and therefore stable, value.

Marshall McLuhan gives an interesting example of the use of leather money, issued in the town of Leiden in the Netherlands when it was under siege by the Spaniards in 1574 – this money was eventually cooked and eaten by the starving population. Another example is the use of packages of cigarettes as currency, after the end of the Second World War.[107]

There are many examples of how monetary value is generated in the art market, something that the graffiti artist Banksy has actively explored in many of his projects. His 'banksy-notes' or 'Di-faced tenners' from 2004[108] are altered (but still regarded as counterfeit) £10 notes with a picture of the late Princess Diana and a few other ludic changes are now fetching prices of about £1000, and one was recently acquired by the British Museum (ironically one of the museums Banksy successfully targeted to place his work in an exhibition as a guerrilla-act in 2005 with a fake piece of rock art – *Peckham Rock* – depicting mimicking a cave-painting with a figure pushing a shopping trolley, and Banksy claims in his book *Wall and Piece* that the work is in permanent collection, and although it wasn't at the time,[109] it was later included in an exhibition on objects of dissent at the British Museum in 2018). The Banksy £10 notes are used by his 'handling service' Pest Control to authenticate Banksy's artworks – they keep half the of the note, and the client gets the other half.

Observed in a semiotic sense, a bank note is more symbolic (or opaque), and the coin is less symbolic (more transparent). This is potentially important for design, when the physical form (or token) has a value based on its replicability (the relation between the type and token).

It is important to note, as Eco does as discussed above, that on principle an exact replica or copy is not possible, so that there is always an element of convention in the interpretation. At the other extreme, the symbolic, we can identify that most financial transactions take place entirely through numbers, that represent monetary value, between banks (or other partners, such as the mechanism of bitcoins and the like).

Another aspect of replicating is the notion of *authenticity*, which is determined by convention. An unauthorised replica of a coin or a bank note is a forgery and if identified as such will not have the same value as the original, although as Eco points out that from a semiotic point of view this doesn't matter, it still refers to the amount of gold it is pretending to represent: "the fact that the bill is a fake merely means that this is a case of lying".[110]

---

* James Bond (Sean Connery) in the 1963 movie *From Russia with Love* has fifty gold sovereigns hidden in his briefcase ("standard kit"), which he offers to his assassin for one last cigarette (and a chance to save his life).

Authenticity in visual art

Umberto Eco brings to this discussion an example from visual art; an exact copy of a classical painting or sculpture is not the real thing and therefore not of the same value. Eco's example is Michelangelo's *Pietà* sculpture (1499), of which the original is in the Vatican, but there are many copies (or replicas) around the world – but this, according to Eco, is not of semiotic concern, "The lust for authenticity is the ideological product of the art market's hidden persuaders."[111] It is worth bearing in mind that many older paintings have been restored, retouched, and repainted throughout their history, so the matter of authenticity is in that sense already a compromise. The recent 'discovery' of a painting attributed to Leonardo da Vinci, *Salvator Mundi* (c. 1500) an example of perhaps too much compromise. Several versions of this image exist, created by Leonardo's students and followers, and while there is enough evidence and opinions that this version is really by Da Vinci, it was sold for $450 million, there are also many doubts (the Louvre reportedly didn't include it in a major Leonardo da Vinci exhibition in 2019). In this case, to me it seems that the restoration included some form of interpretation, perhaps attempting to make the painting look more 'Leonardo-like' or 'Mona Lisa-like'.

Eco mentions Da Vinci's *Mona Lisa* (ca. 1500) in the discussion on doubles.[112] This painting is so iconic that has been used by artists from Duchamp (drawing a moustache on the image, *L. H. O. O. Q*, 1919) to Banksy (his stencilled graffiti tag has Mona Lisa holding a bazooka and wearing headphones, in Soho, London, 2001).[113]

In a work of art that is mechanically or electronically produced (such as silkscreen print, photography, video, and all digital media) the issue of value related to replication is less valid. An early reflection on this notion was the well-known essay by Walter Benjamin, *The Work of Art in the Age of Mechanical Reproduction*, which used the term 'technical reproducibility' – *technischen Reproduzierbarkeit* (originally in German), and particularly looks at the influence on the art work in the transition from theatre to film.[114] This is mainly about recording (and inherently, reproducibility), which had also a far-reaching development in musical practice. In the 1940s the recording medium was used by for instance the French composers Pierre Schaeffer and Pierre Henry, to create music with everyday sounds: *Musique Concrète*.

Value creation and authenticity in contemporary art

Early in the 20th century Marcel Duchamp first explored several painting styles. Particularly Cubism, in for instance the now well-known painting *Nude Descending a Stair Case no. 2* (1912) incorporating not only multiple viewpoints but also a dynamism of time frames, 'dismultiplication',[115] a chronological collation probably influenced by the then emerging medium of the moving image, film, and the chronophotography of Eadweard Muybridge and Étienne-Jules Marey at the end of the 19th century. Duchamp famously started to put everyday objects on display, addressing the replicability, using found objects *(objets trouvés)*. This was certainly not a gimmick, with these 'readymades' Duchamp stated his desire to empha- sise 'art' and not the 'work of art', explicitly addressing the reproducibility: "another aspect of the 'readymade' is its lack of uniqueness ... the replica of a 'readymade' is delivering the same message; in fact nearly every one of the 'readymades' existing today is not an original in the conventional sense" he wrote in 1961. Indeed, the existing 'readymades' are all replicas made in the 1950s and 1960s (authorised by the artist) after the originals such as *Fountain* (1917) (the urinal signed R. Mutt), *Bicycle Wheel* (1913) (mounted on a stool), and *Bottlerack* (1914) (indeed). Duchamp's bicycle wheel has been widely paraphrased and referred to. For instance, it was further developed in 2016 by Shaun Gladwell, whose work often incorpo- rates skateboards and BMX bikes, in the sculpture *Reversed Readymade (Bicycle Wheel)* which "extends the work by returning the functionality of the object". In a later accompanying VR (virtual reality) video a profes- sional BMX rider takes the installation for a spin, and performs some tricks on the unicycle.[116]

Duchamp 'originals' only survive in (mostly casual) photographs of the time, but still the later replicas of the 'readymades' became precious objects, which shows how strongly the art market resists change and per- petuates its own ability to create monetary value. This is also shown in an almost cynical sense in the way that advertising businessman Charles Saatchi monetised his private collection by creating the Young British Artists scene in the 1990s, when I visited the impressive *Sensation* exhi- bition[117] and the Saatchi galleries, with works that often took the read- ymade statement to new extremes (Tracey Emin's bed, Damian Hirst's shark, Gavin Turk's plaque, and Marc Quinn's cast of his head using his own blood – frozen, but also many works that combined strong concepts with impressive craft, such as Ron Mueck's life-like figures, Chris Ofili's paintings (placed on dried elephant dung), and Jake and Dinos Chapman's sinister sculptures). Though all these expressions lacked the replicability of the original readymade idea of Duchamp, they showed that monetary value can be generated artificially (if there is enough artistic merit) – in the decade afterwards, many of the works were sold on for multiple millions of dollars (each).

Prints of a work by graffiti artist Banksy from 2006, mocking the art market with a scene of an auction where people are bidding on an artwork with the text "I can't believe that you morons actually buy this shit", actu- ally sell for tens of thousands of dollars.[118]

In a recent event Banksy took things even further: one of his framed prints partially self-destructed at an auction; after the initial shock, it turned out that this immediately increased its value! The print was mounted in a large frame, which started to shred the print when the auction hammer fell. There have been self-destructing art works before of course. Most notably kinetic artist Jean Tinguely, who created self-destructing sculptures from *Hommage à New York* in 1960 to *La Vittoria* in Milan in 1970.[119] There have been other acts of destruction, such as pop musicians Bill Drummond and Jimmy Cauty (of the band KLF), who always had a strong conceptual bend (some of their music was pure *musique concrète*, and their video clips often unorthodox); in their desire to be seen as artists performed an indeed impressive artistic and conceptual (albeit costly) act of burning £1m in bank notes in 1994. In 2016 Mike Parr, best known for his artworks and performances involving self-mutilation, publicly burned over 30 m² of his drawings of an estimated worth of $750k.[120] Banksy went further than this, first of all the print *Girl with Balloon* was just that, a print but authenticated by Banksy's entity Pest Control as genuine, which made it worth bidding up to nearly £1m for it. In its half-shredded form the print is now worth much more, not only in monetary value but as a unique artistic achievement. The act went viral, several people who didn't even know who Banksy was told me about it at the time, but as Patrick Potter writes in an essay *Mister Self Destruct – shredded art* this is "the central paradox – that attacks on the hypocrisy of the art market revitalise the art market".[121] The importance of the fact that the Banksy print was original (though "a much replicated original"), shows how the work of art has shifted away from the importance of initial uniqueness to a mass-produced uniqueness, quite a feat of the art market.

It follows the tradition of Andy Warhol in the 1960s, who used silkscreen printing techniques to create multiples of artworks, often depicting popular images (such as the Campbell soup cans, and he too had to wait more than a decade before they became worth millions), and other pop artists at the time such as Roy Lichtenstein's prints which were based on cartoon graphics style.

Since industrialisation most objects around us are mass produced, which gives the hand-made and the unique a particular value. This relationship is challenged by the role of instant manufacturing (techniques such as 3D printing) on the value, and potential for increased relevance, of individually produced objects and environments. There is more on this issue in the discussion of the role of instant manufacturing on the value (and potential for increased relevance) of individually produced objects and environments in Chapter 12.

LANGUAGE AND THOUGHT

Signs can also represent non-physical objects, such as a process (the weather, flow) or an idea. In addition to nouns (things), verbs (actions) can be represented. The actions can have different degrees of physicality which will result in a different degree of likeliness in the iconic representation. The power of the sign is that it can also be used for representing completely abstract notions and concepts, which allows for reasoning and discussing (and writing books ...) to advance knowledge and insights. It is a way of externalising our thoughts, which then can be re-internalised with reflection. This powerful mechanism is the basis for human development.

De Saussure and others even insisted that *all* thought is in language, and dialogic. Peirce for instance has stated that "every thought is a sign", and that "All thinking is dialogic in form. Your self of one instant appeals to your deeper self for his assent", and that even ideas are signs.[122] This seems true for rational thought, which can take place through this kind of role-playing in the mind, facilitating the inner dialogue which can be the thinking that we are aware of, but the mind is more complex than that – often thoughts and ideas are non-verbal, 'thought flashes', which seem to well up with an enormous complexity, multileveled and simultaneous. This is sometimes called 'mentalese' or 'brainish'.[123]

Dialogic rational thinking is a process of shaping and mental articulation in language. The next step is the verbal or gestural (or mediated through typing) external expression, further articulation and modification can take place in this process. All these processes can be interrelated, and influence (enhance, modify) the outcomes of the process – also negatively, one can reason until a patently bad idea is fixated and becomes seen as a reasonable course of action. This is further discussed in the context of unconscious processing in Chapter 10 *Thinking*, and in Chapter 11 *Expressing*. But at this point it is a good context to look at some examples of different types of language, and how they might influence thinking – or not.

Language and time

There have been many misconceptions about how language influences our thinking, particularly in the first half of the 20th century when Edward Sapir, and later his student Benjamin Lee Whorff, attempted to link insights from anthropological studies into a range of cultures, to language. This came from a fascination, appreciation, and inspiration (but not always comprehension) of different ways of thinking (or being) that different cultures can have. Reflecting the general attitudes of the time about race, theories like the Sapir-Whorff hypothesis were developed that tried to explain a hierarchical view of cultures, from 'primitive' to the more sophisticated. The linguist Guy Deutscher, in his book *Through the Language Glass – why the world looks different in other languages* gives a historical overview in the chapter 'Crying Whorff',[124] and actually devotes most of his book to show the opposite of what his title suggests. For instance, if a language doesn't have a word to describe a certain concept, it doesn't mean that a person speaking that language can't think about such a concept.

Daniel Everett gives examples of this too, from the Pirahã people in the Amazon (mentioned in Chapters 2 and 3) who don't use numbers (because they don't need them, as Everett emphasises, they use words that roughly correspond to our one/two/three, or rather, one/a few/a lot). This doesn't mean for instance that a parent can't know how many children they have in a large family.[125]

Another common example from Whorff's writings was that the Hopi people (the Native American Pueblo people living in what is now Arizona*) wouldn't have a concept of time, as their language supposedly didn't have words for time. This turned out to be not quite true, as in the 1980s it became clear that not only does the Hopi language have words for time, it is mainly that their concept of time is *different*, not absent.[126] Anthropologist E. T. Hall in *The Dance of Life* [1983] is more nuanced and certainly respectful about the Hopi people, although he holds Whorff in high esteem. Hall actually lived and worked with the Hopi for years in the 1930s in order to learn about their language and culture. Hall realised that in Hopi time there is a strong link between past, present, and future, a common notion in Indigenous cultures, a cyclical and interconnected chronology as opposed to the spatial, ordered notion of Western society.[†] This is why I always have been sceptical of approaching time as a 'fourth dimension', from which the notion of time travel seems acceptable (in spite of all its obvious paradoxes).[127] Hall has studied many cultures, and although sometimes a bit generalising, he does have a point when he describes polychronic and monochronic cultures. Hall sees some northern cultures (in Europe, America, Japan) as monochronic, doing one thing at a time which leads to high punctuality. Polychronic cultures, such as in Mediterranean countries

---

* There is one word from the Hopi language that has become very famous, *Koyaanisqatsi* ('life out of balance'), the title of the abstract film from 1982 of director Godfrey Reggio, cinematographer Ron Fricke, and composer Philip Glass. The film is showing how crazy Western life actually is (in the imagery of 1980's US, mostly); the film is a meditation on the relationship between society and technology, and the cinematography, editing, and music are all driving a non-linear narrative. Glass's minimal music patterns are the basis of most of the editing and imagery, which is unusual in cinema. Later examples are *Lola Rennt* (Run Lola Run), the 1989 film by Tom Twyker (which also explores cyclical time in a repeating yet shifting narrative) and is driven by the music, and *Baby Driver* from 2017 which is based on musical tracks mostly chosen by director Edgar Wright, working with composer Steven Price and sound designers and editors; the movie is driven entirely by the music.

† Ironically the notion of cyclical (or dialectic) time is also a key aspect in the narrative of the 2016 science fiction movie *Arrival*, where a very advanced alien species approach Earth to help us with this insight, giving us 'the gift of dialectical time'. Though a further irony is that the film perpetuates the Sapir-Wolff hypothesis, that by learning the alien's language we would also magically acquire this view of time and become clairvoyant. However, the language of the aliens shows a number of important aspects, such as their spoken language not being related to their written language (all human written languages are based on spoken language), and the (for me) most striking aspect of all-in-one utterances, the circles they produce as sentences are expressed instantaneously (unlike our written languages which are all produced serially).

(and also the Hopi people), are more about doing multiple things at the same time, less planned, potentially more appropriate certainly in complex situations (such as writing a book that spans many disciplines and approaches ... needless to say, personally I feel more affinity with the polychronic mindset) – things are finished when they are finished.

The notion of past, present and future time as linked can be seen in the Quiché culture in what is now Guatemala. The Quiché culture descended from the Mayans, and has two meshing calendars – one of 365 days for civic purposes, and one for religious purposes of 260 days.[128] This calendrical system is like a wheel, without a beginning or end, unlike the calendar common in the Western world. The calendars play an important role in organising knowledge, so that it can be remembered.[129]

Deutscher gives an example of the other extreme, of the language of the Matsés people in the rainforest of Peru, in which there are up to three versions of past tense (in English and most Western languages there is only one).[130] Speaking about a past event in this language, one has to specify if something happened in the recent past (in the last month), less recent (from one month to about 50 years), or the remote past (more than 50 years). In the Matsés language one also has to be very specific about modality or evidentiality, distinguishing between whether an event has been witnessed first-hand, or inferred from evidence, or assumed from habit (conjecture, an event which is known to be likely to have happened), or reported by someone else. Of course, as Deutscher points out, all these distinctions can also be made in English (or his native German), but the main difference is in this example that the Matsés language forces the speaker to be this specific.[131] There are Australian Aboriginal languages in which this also occurs, such as Dyirbal, where the speaker has to be explicit about what one knows and how one got to know it (rather than having a generic verb for to know).[132]

Language and colour

Another common misconception is that if a language has more colour words, it is more sophisticated (and by extension, so are the speakers of that language). Indeed I recall being somewhat impressed when only recently I learned about the colour 'teal' in English, a kind of blue-green (such as cyan and turquoise) for which we don't have a word in Dutch (*taling* is just a kind of duck, not a colour, we do have *cyaan* and *turkoois* but they are different shades in the blue-green spectrum), but indeed I do perceive this colour as distinct even without having a word for it (though I actually don't really like this particular colour).

In the evolution of a language, according to Deutscher, there is an order of names 'invented' for colours over time; after black and white the first colour in most languages that was named was red.[133] This may seem surprising, we are surrounded by vast areas of other colours, blue sky or sea, green leaves, brown soil, but that is presumably precisely why we had less of a need to label these – if it is around us in abundance there is not much reason to mention it. The colour red – blood! fire! danger! – is really worth talking about.*

The next colours named in most languages were yellow and green (fruit colours, as Deutscher reports, the important difference between ripe and unripe), and finally blue. Some of this is related to physiological factors, the wavelengths that the respective receptor types in the human eye are sensitive to. This is discussed in more detail in the section on the visual sense in Chapter 8, *Feeling*. (There is a funny coincidence that the first LEDs (Light Emitting Diodes) that were widely available in the 1970s were red ones, followed by yellow and green ones, and only much later in the 1990s (and for a long time much more expensive) the blue ones.)

The insight that languages as they evolve develop more colour words over time, may have led people to believe that the number of colour words is a measure for the sophistication of a language. Although there are examples where language or vocabulary influences the way we see the world, in *Through the Language Glass* Deutscher shows that this is mostly anecdotal and subjective, and not nearly as strong that one could claim that "the world looks different in other languages" as the subtitle asserts. The only clear objective proof that language can have an influence on thought is shown in a number of experiments where participants perform a choice reaction test with (shades of) colour. After all, colour names are actually kind of arbitrary, because the colour spectrum is continuous, so the 'boundaries' between those named colours are arbitrary. Confusion around the boundary between green and blue (or teal) is a particularly popular discussion topic, also because we can't know what the subjective experience of an individual is (the domain of phenomenology[134], further discussed in Chapter 9 *Experiencing*). The choice reaction tests are the topic of the last chapter in Deutscher's book, where he reports projects carried out by colour researchers. In these types of tests, participants are not asked about their subjective experience of a colour but asked to match (shades of) colours. Their reaction times depend on the distance in shade between the target colours (if they are close together, it is harder to distinguish and the reaction time is slower, when the shades are further apart and the task becomes easier). But the reaction times are also influenced by the names that our language has assigned to them. This is shown in one test which is set up with two conditions, one forcing the participant to use their right eye (and therefore the left hemisphere of their brain, where most of the language faculties are thought to reside – such as Broca's area), or their left eye (and the corresponding right hemisphere). In the former case the reaction times are influenced by the language labels.[135] Further research using MRI (magnetic

---

* Maybe this is why it is the forbidden colour in the movie *The Village* of M. Night Shyamalan from 2004, every occurrence of anything red is immediately eradicated.

resonance imaging) brain scanning also supports the finding that the colour names influence the reaction tasks corresponding with observed activity of the language centre in the brain.[136] But the most striking research compares English speakers with Russian speakers.[137] In Russian there are separate words for light blue, 'goluboy' and dark blue, 'siniy', a good example if something is familiar it doesn't mean it has to make sense. I checked this in 2011 with two of my master students from Russia, they confirmed this and actually seemed to think that it is strange that in English we don't make this distinction, and that we call two different colours by the same name! (This is even more fun than discussing whether something is blue or green – or teal.) Indeed, while my favourite colour has always been blue, and there are some particular blues that I really like, it is across the spectrum and not confined to 'siniy' or 'goluboy'. Similarly, blue is a popular colour in the graphical interface, many of Apple's and other applications icons are blue (making them hard to distinguish), and the whole visual interface that Victor Donker designed for the Stepping Tiles project[138] is ... blue (a mixture of 'siniy' and 'goluboy', actually).*

In the test, participants were asked to match shades of blue, and it turned out that the reaction time was not only influenced by the distance in shade as expected, but in the case of the Russian speakers also around the boundary between goluboy and siniy! This effect disappeared when the participants were asked to engage in a parallel verbal task, to engage their language faculty away from the colour shade recognition task, proving that it really is a linguistic influence.

While this work is by some seen as 'neo-Whorff', there is nothing wrong with looking at differences in thinking relating to language, provided there is no value judgement involved.

Language and orientation

Another interesting and particularly striking example of how language can influence thinking, and the way of being in the world, is of cultures where it is common to indicate direction in compass terms. Instead of the system which is most common to us, of egocentric referencing – not in the sense of egotism, but by using phrases such as 'to my left' or 'behind me'. By contrast, in a geographic directional system, one would use phrases such as 'to the east', 'to the south-west of my foot'. This means one has to be continuously aware of where north is, which is something most people living in cities nowadays are less aware off. Giving people directions based on the compass is often seen as confusing, along with the increasing reliance on GPS navigation systems which can display the map in egocentric

---

* One of Google's April Fool's jokes in 2013 was "introducing Gmail Blue". The same Gmail, but 'blue-er'.

orientation (instead of the traditional convention of north being the top of the map). Some cities are easier to navigate in geographic terms, if they have a particular landmark building (Paris, New York City), a particular clear town plan at least in certain areas, oriented on the cardinal directions of the compass (Paris, Barcelona's Eixample, Manhattan in New York City) (Amsterdam's centre is clearly laid out with its semi-circular canals ('de grachtengordel'), but because of this bent structure actually hard to keep compass direction.), a main river, bay, or sea front (London, Paris, Sydney, The Hague), or mountains. Indeed, in places where the whole area is oriented on a slope of hill or (volcanic) mountain, it is even used in everyday language to reference direction as "uphill" or "downhill" (I experienced this in Bali). These properties of the physical environment, layout, orientation, etc., whether human-made or natural, can play a prominent role in navigating. It is another form of scaffolding, of course, studied in urban design, town planning, and landscape architecture.

The system of cardinal referencing is used by for instance the Guugu Yimidhirr people, in what is now northern Queensland, near Cooktown (I visited the area in 2012 when participating in an Indigenous bushfire management camp, as discussed at the end of Chapter 4) and researched since the 1970s by the anthropologist John Haviland.[139] There is one word of their language that everyone knows: 'kangaroo'. This is a generic word in the English language, based on the Guugu Yimidhirr word for a specific species of marsupial, *gangurru*, as told to Captain Cook when he spent time in the area on his voyage exploring the Australian east coast in 1770.[140]

Ever since I read about the Guugu Yimidhirr people's orientation skills in Deutscher's *Through the Language Glass* in 2011, I tried to be more aware of the cardinal directions, by regularly practising, something that was already necessary due to the otherwise subconscious influence of the position of the sun in the southern hemisphere (in the north, instead of the south). It took me a year, through practising conscious awareness, to internalise the sun direction. Often during presentations, I ask the audience members to try and point north with their eyes closed (to avoid copying each other), which has led to a nice collection of photographs of people pointing in all possible directions. Particularly in underground lecture theatres this is very difficult for the untrained, but for the Guugu Yimidhirr people this is no effort at all. They can be led into a building blindfolded and taken through labyrinth structures and still know where north is. There has been speculation of an actual sensitivity to the earth magnetic field[141] but this has not been proven. I think this shows a magnetic sensory modality is not necessary, that this is an ability that emerges from habit since birth, which scaffolds this ability, similar to the advanced ability to discriminate pitch by a trained musician (some even have 'perfect pitch', which is an absolute sense, and Deutscher likens the 'compass sense' to 'perfect pitch for directions'[142]).

Categorisations

We are very good at generalising, recognising objects and images as belonging to the same category: coins, cats, handwritten words, trees, etc. We are pattern recognisers, as Marshall McLuhan often observed, and good at generalising into categories. Some of this may be innate, but many of our pattern-recognising abilities need to be learned from birth by associating the recognised similarities, eventually putting them in categories with symbolic labels such as 'trees', 'fruit', 'nuts', 'bottlecaps', etc. It has been pointed out that this is a typical habit of Western culture, while this is less so in traditional (oral) cultures. For instance in Australian Aboriginal languages there is not a generic word for 'tree' but each tree type is identified by its specific name.[143] This of course is also related to the fact that the traditional Aboriginal culture has a much closer relationship with the natural environment than the Western culture; these are details that matter, but it is interesting to note that in traditional Aboriginal culture one has to be specific about matters that in Western culture are generalised.

In the documentary about the making of *Ten Canoes,* the 2006 movie written and played by Indigenous actors in the Yolngu Matha language from northeast Arnhem land, director Rolf de Heer reflects on languages: "Ours is a language of classification and categorisation. Theirs is a language of connection and unity. Everything is all one." This also refers to the notion of time in many Indigenous cultures, where past, present, and future is seen as one, as mentioned above.[144]

As linguist Robert Dixon points out, in the past this lack of generalisation was seen as an inability to think in the abstract, but in fact it is a strength, a sign that language is used to be precise and specific where appropriate. Being vague is seen as "a mark of foolishness, or lack of brain".[145] In Australian Aboriginal languages, such as the Dyirbal language that Dixon discusses, there are actually two versions of the language, the Guwal everyday language (in which one has to be specific) and the Jalnguy style which is used in polite or taboo situations. It is common in Australian Aboriginal cultures to restrict interaction based on taboos, particularly with one's mother-in-law.[146] In Jalnguy more generic terms are used, in what is also known as 'avoidance style'.[147]

A large part of Umberto Eco's book *Kant and the Platypus – Essays on Language and Cognition* [1999] is concerned with categorisations. Eco makes a distinction between what things look like, or seem like (the Nuclear Content or NC) and what they actually are (the Molar Content, or MC). This relates to a definition as found in a dictionary (representing the meaning of a term, not organised in a structure, as such it could even be assembled in a bricolage style), and an encyclopaedia (the meaning of a term in a broader context, in relation to other content, organised in a structure).[148] Hence the platypus in the title of the book, an animal which actively defied categorisation in the existing taxonomy, being a combination of mammal (it has milk glands, fur, and four legs), and a bird (it seemed to have a beak and lays eggs), and lives under water. Because of this challenge to the taxonomy (MC), there was initial confusion about the nature of the animal (after its discovery in Australia early in the 19th century, its actual existence was initially doubted by observers back in England). In addition to the elusive nature of the platypus, due to the tension with the categorical framework in use, it took over 80 years since the animal was discovered before an actual platypus egg was identified (in 1884) but by then it stayed in its designated place in the category of mammals (as an NC relic).[149] The encyclopaedic interpretation would take into account the notion of evolution, and the importance of the discovery of the platypus (and the echidna) in supporting this theory. After all, the platypus has evolved from a bird, not from mammals, as discussed in Chapter 3 in the section on evolution.

This tension is also illustrated by the case of dolphins and whales, which seem to be fish but are mammals (as everyone now knows). Another example that Eco uses is that of Uluru in the middle of the Australian continent.[150] Eco explains that Uluru looks like a mountain and is best described as such if one were to give directions to find it in the desert (a dictionary description), but according to Eco it is in fact a rock (it is a large 'monolith planted in the ground') (an encyclopaedic description).

PHATIC COMMUNICATION AND INTERACTION

Not all communication is about exchanging information and meaning. The phatic function is about social communication, rather than meaning, such as small talk, gossip and 'chit-chat', and also sometimes described as 'testing the channel'. Indeed, the term is derived from the Greek *phatikos*, affirming, or *phatos*, feeling, which is also where the word empathy comes from, 'feeling in', *invoelen* in Dutch.

This phatic communication is an interesting concept, and although of course an ancient form of communication, it was first formulated as such in the 1920s by anthropologist Bronislaw Malinowski (then a professor at the University of London), in an essay published as a 'supplement' in Ogden and Richards's book *The Meaning of Meaning* in 1923. In phatic communication, Malinowski wrote, words "fulfil a social function and that is their principal aim", and that "language does not function here as a means of transmission of thought".[151] Phatic communication is sometimes referred to as 'social grooming', and indeed comparable with grooming rituals between animals. Grooming always was a physical ritual, of attending each other's coat, cleaning fur and skin particularly of parasitic insects, and it also had a social bonding aspect including reconciliation after a conflict.[152] David Linden describes the role of grooming at the beginning of his book *Touch*, in the chapter that discusses "the skin as a social organ". Linden gives many examples of how people exchange information through touch, often unconsciously, and the origins in animal grooming.[153] As mentioned in Chapter 3, due to the increase in typical group size, as well as (later) the loss of hair, early humans may have developed a vocal version of this type of grooming communication, in order to maintain these social connections.[154] This vocal grooming can be seen as a compensation for the loss of the function of physical connection through grooming. Dan Everett writes about this too, from his experience of living with the Pirahã people in the Amazon. They have less need for vocal phatic communication as they also use physical grooming, and live in small communities, though this doesn't entirely explain their lack of 'small talk'.[155]

Small talk and chit-chat

Phatic communication is very common in our interactions, to set up (or avoid) a deeper conversation. Phrases such as *ça va?* in French, *hoe gaat het?* in Dutch, and *how are you?* in English are usual starting points, to which one responds with something like *ça va bien, et vous?*, *gaat je niks an!*, and *good, and you?*, respectively. These aren't really questions of an inquisitive nature, rather of the rhetorical kind, and depending on the local culture. As a Dutch person, used to quite succinct exchanges of this type, I had to get used to the generous Australian vernacular, and I quickly learned not to elaborate in response to *hi mate, how's your day being?* with too many details about my actual experiences and impressions of the day so far, instead learned the correct phatic response in this situation would be something like *alright*, or even *not too shabby, yourself?*

Another common example of phatic communication is dialogues about the weather, tedious to the outsider but essential as a form of establishing the connection between the participants in the communication. The importance of this is shown by the fact that even in Sydney, where the weather is sunny and pleasant most of the time, people still talk about the weather a lot! But it seems that any topic will do, from gun laws to barnyard animals, or sinterklaas folklore. And it is an important element in social communication; a recent book in the Netherlands gives practical guidance in how to "survive small talk" and make effective use of it.[156]

We still have a large repertoire of types of physical communications of this phatic kind, largely unconscious as Linden points out, but often deliberate. The handshake, kiss on the cheek, the embrace, are all essential first steps towards developing strong bonds and relationships, and these steps should not be skipped. The handshake not only has the potential to reveal a number of personal characteristics to the receiver, it also gently enforces everyone in a group encounter to have a personal exchange, however briefly. This is important for establishing social bonds – the handshake cannot be broadcast, it is always personal.

Fidgeting

There is also a physical version of phatic communication, in the way we can enjoy the interaction with objects and devices. This occurs particularly in mechanical technology, and in the mechanical aspects of electrical and electronic technologies. This kind of 'phatic interaction' is found in the joy of fiddling and fidgeting, of manipulating handles and levers (mechanical systems), or switches and dials (of electrical and electronic systems). It is therefore a pity that our current digital technologies, such as computers, tablets, and smartphones are increasingly losing the physical and mechanical controls – Apple as usual leads the way, in this case in the wrong direction, by removing physical controls or limiting them (volume up/down buttons instead of a dial, and the recent iPhones have lost the 'home button' entirely, which already had way too many functions, and which is now integrated into the screen. And although the interaction bandwidth of the touchscreen has increased over the years, it is still a loss that the physical controls are gone. Manufacturers avoid physical controls because they are more expensive, can assemble dirt and moisture, and have a limited life span, but I argue that it is worth it. On the other hand, Apple has also developed the Taptic Engine which can be found in their iPhones since the 6s model and in the trackpad on the MacBooks, a kind of solenoid which creates the feeling of a (virtual) tap – much more pleasant than the more common buzzers based on a spinning motor with an eccentric weight (vibromotor) which is mainly useful for alerting. In fact it is so pleasant that I sometimes fiddle with it, turning the iPhone to silent (a switch!) activates Taptic Engine and it generates a nice double tap.

The increased reliance on the touchscreen for all interaction has led to a range of products which have no other function than to be fiddled with, not just for people with ADHD, many people seem to enjoy fidget spinners, fidget cubes, Tech Decks, Rubik's Cubes, and the recently trending Pop It bubble toys. (Though the COVID-19 pandemic has led to an increase of the emphasis on non-physical interaction, unfortunately.)

AFFORDANCES AND DIRECT PERCEPTION

It is important to note that the field of semiotics is mostly concerned with communication and language between people. Particularly the early phase of semiotic study had a strong emphasis on human and animal language and communication, and less on physical objects and how meaning relates to this interaction with the physical world. As Umberto Eco notes in *A Theory of Semiotics*: "Objects are not considered within Saussure's linguistics and are considered within Peirce's theoretical frameworks only when discussing particular types of signs such as icons and indices."[157]

Semiotics is a vast, multifaceted, and complex field, precisely because it deals with human language and social behaviour. I have tried to capture from semiotics what I think is most relevant for the context of understanding interaction, including the physical, because interaction is not only about language but also about manipulation. Semiotics is mostly about signs, generally made to represent meaning. Objects and environments have a presence of their own, and have a potential for a direct relationship with the human or other animal. This is known as affordances, which can play a role in the relationship between the object or environment, and the animal. Affordances can reveal to the observer what potential actions and behaviours are possible. The term is increasingly used in other contexts such as media theory, where it is used as a kind of synonym for 'functionality'. This is not necessarily wrong of course, but it erodes the term affordance to meaning any function of an object, which is a pity as we really needed a specific term for what actions and behaviours an object can potentially offer, though its affordances.

The concept of affordances is discussed in great detail in Chapter 9 *Experiencing*, and in particular, the ecological approach and the notion of direct perception.

Ecological approach to perception

Not all of the information in our environment will be recognised as such by all perceivers, it depends on their individual knowledge, experience, approaches, values, and mindset. This reciprocal relationship between people (or any animal) and their environment is the basis of the ecological approach to perception that the psychologist J. J. Gibson developed from the 1960s. Going against the then current notion in cognitive science of perception as information processing and inference, Gibson showed that information in the environment can be picked up through direct perception of objects and the environment, particularly their *affordances*. Affordances are properties of the objects to do with their potential for use, which relates in different ways to different animals (including people). Gibson introduced the term in his second book in the 1960s, but the notion is present in his earlier writings from the 1950s, and worked it out in full in the 1970s.[158] Gibson explains how an object or entity can present potential for interaction and meaning through its mere presence, and how this idea was already present in the ideas of the Gestalt psychologists in the early 20th century. Affordances are an important notion in design for interaction, and design in general – designers have always been involved in supporting intuitive and direct perception of the potential functions of the artefacts.

The notion of affordances was introduced to design by cognitive psychologist Donald Norman, in his insightful book *The Psychology of Everyday Things*, which became a bestseller after it was renamed *The Design of*

Everyday Things in the 1990s.[159] Unfortunately Norman didn't accept or support Gibson's ecological approach, nor the notion of 'direct perception', as Norman explains in many publications since, limiting his interpretation of affordances to 'perceived affordance'. Designers however enthusiastically embraced the use of the term affordance, and direct perception – because this is what designers have always been doing. It makes perfect sense, whether designing a product, layout of a page, clothing, buildings – a good designer always tries to make the design as clear as possible, so that it can be interacted with in an intuitive and inherent way, 'directly perceived' without needing conscious mental effort of processing information. Recent insights on the role of unconscious thinking as a massively parallel and fast process in resonance with the environment further support the notion of direct perception.[160] Designers have always known that a design (layout of controls for instance) can be perceived differently by different people, depending on their knowledge, abilities, behaviours, and needs, just like the 'interpretant' in Peirce's framework, and that the 'perceived affordance' is not a fixed thing. Gibson's ecological approach therefore makes sense to designers.

Meanwhile, Norman has proposed to label affordances (or his version of it, 'perceived affordances') as *signifiers*. In a way, this is good because now we have clarity, Gibson's (and designers') affordances are potentially directly perceived, and Norman's (and other cognitive scientists') signifiers need processing. This choice of label is a bit unfortunate of course, because as has been discussed from the beginning of this chapter, in the field of semiotics and linguistics, 'signifier' is the physical presence of a sign, that communicates the meaning, the 'signified', as defined originally by De Saussure. Affordances are not so much about meaning but about potential for behaviour and action – and so are 'Norman-signifiers'.

A sign *represents*, it *stands for* something, while an affordance *presents*, it *is for* something.

Affordances and semiotics

The notion that an object can 'radiate' meaning and usage potential is quite straightforward when viewed in an ecological approach. It also fits nicely with the core concept of semiotics that *meaning emerges*, between participants in the interaction. Affordances are discussed in semiotic literature, but it is rare. To many semioticians *everything is a sign*, whether intentional or not, natural or artificial, real or virtual. Most of the field of semiotics is concerned with the way meaning is conveyed through coded information, but some approach this from observations of the physical environment. In Peirce's framework and terminology as discussed above, affordances can be seen as related to his concept of 'firstness' of an object, and also in some of the reflections of Sebeok when he looked at animal and environment interactions (drawing on the then current notion of the *Umwelt* from biology).

The strongest connection can be found in the work of Umberto Eco. He described affordances (but without using the term) in *A Theory of Semiotics* in 1976:

> Signs are also distinguished according to their *semiotic specificity.* Some signs are objects explicitly produced in order to signify, other are objects produced to perform a given function. These latter can only be assumed as signs in one of two ways; either they are chosen as representatives of a class of objects or they are recognized as forms that elicit or permit a given function precisely because their shape suggests (and therefore 'means', 'signifies') that possible function. In this second case they are used as functional objects only when, and only because, they are decoded as signs. There is a difference (as regards sign-specificity) between the injunction /sit down!/ and the physical form of a chair which permits and induces certain functions (among others, that of sitting down); but it is equally clear that they can be viewed under the same semiotic profile.[161]

The affordance concept is often illustrated with the example of the chair, for instance, Gibson wrote "If a surface of support [...] is [...] knee-high above the ground, it affords sitting on." It is the affordance of "sit-on-able". And to illustrate the relational and ecological aspect of affordances, he continues "Knee-high for a child is not the same as knee-high for an adult, so the affordance is relative to the size of the individual." A child would probably climb on it, which is a different affordance, that of "climb-on-able".[162]

In *Kant and the Platypus* in 1999, Eco does use the term affordance and discusses the notion further. Here he referred to a French author (Luis Prieto) who uses the term *pertinence*.[163] Eco also discusses the influence between the modes: the name given to an object (symbolic, abstract label) can influence the (direct) perception of the function of the object.

An object can of course have multiple modes of communication simultaneously. It can have affordances, which potentially reveal opportunities for use, and behaviours. But it can also be a sign, by mimesis or by convention, particularly dependent on the cultural context. Wearing a bike helmet in Sydney not only makes use of the affordance of protection for the wearer, but it also signals to others that the wearer is mindful of safety (whereas in the Netherlands wearing a helmet is uncommon, so the signalling function is different). Another example is the face mask, which not only affords some level of protection to microbes for the wearer, but also signals that the wearer is being cautious and responsible. During the Covid pandemic this soon became a statement, including people deliberately *not* wearing a mask to express certain beliefs.

The affordance of affordances

In Chapter 9 *Experiencing* (in Volume 2 of *Understanding Interaction*) affordances are discussed in depth and a broader context, and a framework is proposed: 'the affordance of affordances'. The framework is based on a large range of literature (from psychology, design, semiotics, media studies, art theory), other research, contemporary similar approaches such as enactivism and sensorimotor theory, and some of my recent explorations and experiments.

The 'affordance of affordances' framework extends the vocabulary by adding appropriate adjectives to the noun affordance. This indicates particular attributes of affordances such as: obvious, obscure, appropriated, intentional, deceiving ('twisted' affordance), false ('fake affordance'), inhibiting ('un-affordance'), dynamic, altering (changing the affordance, an 'affordifier'), positive, negative, inviting, enforcing, accidental, serendipitous, etc.

The role of an affordance in the manipulative mode can be compared with the role of a written instruction in the symbolic mode, or the function of a pictogram in the iconic mode (and all of these can occur simultaneously, complementarily, and redundantly).

NOTES

1. A good overview of the theories of developmental stages in children's learning is *How Children Think and Learn – the social contexts of cognitive development* by David Wood [1998].
2. [Diamond 1997, p306].
3. [2019, p4]. Dixon authored several books on Australian languages, including dictionaries. See also *The Little Red Yellow Black Book* [AIATSIS and Pascoe 2018, p42].
4. Shannon had published a paper earlier on the topic with the same title (in *The Bell System Technical Journal*, July and October issues in 1948), which was later expanded with other essays by Weaver into a book [1949]. Shannon's revolutionary ideas and his model are discussed in detail by James Gleick in *The Information* in Chapter 7, 'Information Theory' and Chapter 8, 'The Informational Turn'. [2011, pp204–268]. Shannon's model is presented widely in other publications in a range of disciplines; for instance in semiotics [Eco 1976, p33, p141], media theory [Mulder 2004, p31], and semantics [Lyons 1977, pp36–37]. See also Tor Nørretranders discussion of Shannon's ideas [1998, pp96-103].
5. [Gleick 2011, p409]. Also famously proclaimed by Frank Zappa on one of the *Joe's Garage* albums in 1979, as part of a slightly different hierarchy, thank you Jos Mulder for reminding me.
6. [Kelly 2016, 2019].
7. [Gleick 2011, p222].
8. [Shannon and Weaver 1949, p31].
9. [Nørretranders 1998, p40]. Danish science writer Tor Nørretranders was one of the first to write in-depth about how the human mind (re-)creates a

sense of the real world inside our minds from a contemporary physics point of view, at the same time as Daniel Dennett's *Consciousness Explained* [1991]. Nørretranders in the first chapters takes an in-depth look into physics, particularly the notion of entropy (and the 'thought experiments' that proposed ways to re-ordering disorder and thus refuting this second law of thermodynamics), and the analogies with human (un)consciousness and the notion of information entropy. This is reflected in the original title of the book from 1991(!), which is about being in touch or making a mark on the world. In the English translation the title is *The User Illusion*, which is an HCI term (the user's mental model); Nørretranders writes in a chapter with that name about how he was inspired by the work of Alan Kay and his colleagues at Xerox PARC working on first graphical user interfaces in the 1970s and 1980s [1998, pp290-293](as discussed in Chapter 1 *Interacting*). Nørretranders refers to an article in the *Scientific American* by Kay where the concept of the 'user illusion' is introduced [1984]. Towards the end of the article, Kay describes the user illusion as 'theater, the ultimate mirror' [Kay 1984, p58].

The subtitle *cutting consciousness to size* is more apt. The original title of the book in Danish is *Mærk Verden – En beretning om bevidsthed*, and I asked my friend Kristina Andersen from Denmark, formerly at the STEIM studio in Amsterdam, now at the Department of Industrial Design at the TU Eindhoven in the Netherlands, about the book and she responded (by email on 30 June 2020): "The book was a big deal in DK when it came out, I remember that I attended a big lecture he gave and he was on TV a lot. The title literally means: *Feel (or touch) the world*." (My italics). I think that this is an appropriate title, it is about how we are (or think we are) in touch with our environment. Or as Nørretranders expresses it in several points of the book, 'making a mark on the world'. This is further discussed in Chapter 9 *Thinking*, which is about how (we think) we think. It also links back to ideas of René Descartes early in the 17th century, as covered in the section *Science – reason and knowledge* at the beginning of Chapter 5 *Industrialising*.

10. [Shannon and Weaver 1949, p9, p32].
11. [Gleick 2011, p4, p229].
12. *Over Mediatheorie – taal, beeld, geluid, gedrag* [Mulder 2004, p32].
13. [Gleick 2011, p247].
14. [Gleick 2011, p204].
15. As mentioned in Chapter 1. [Gleick 2011, p238], [Wardrip-Fruin and Montfort (eds.) 2003, p65].
16. [Nørretranders 1998, p42].
17. [Shannon and Weaver 1949, p28].
18. [Hall 1983, pp59–63]. Edward Hall also wrote about *proxemics*, emphasising the relevance of physical distance in social communication, in *The Hidden Dimension* [1966]. This is further discussed in Chapter 11 *Expressing*.
19. [Lyons 1977, p4].
20. See for instance the two volumes on semantics by John Lyons [1977]. Professor Lyons later became Master at Trinity Hall in Cambridge, which was my college when I was a researcher there in 1999–2000, so I might have caught a glimpse of him but at the time I barely knew anything about semantics, semiotics, or linguistics. Sir Lyons retired from Trinity Hall in 2000, and passed away in 2020.

21. [Eco 1976, p15], [Cobley and Jansz 1997 p21], [Ogden and Richards 1923, p280], [Hodge and Kress 1988, p20].
22. [Gibson 1966, 1979].
23. Excerpts of the translation of Ferdinand De Saussure's text are presented as an essay 'On the Nature of Language' [Lane (ed.) 1970, pp43–56]. I prefer to refer to him as 'De Saussure', though it is common in semiotic and linguistic literature to refer to 'Saussure'.
24. [Hodge and Kress 1988, pp15–18].
25. [Cobley and Jansz 1997, p38], [Hodge and Kress 1988, pp18–20].
26. I have used a later version, the eighth edition [Ogden and Richards 1946].
27. A section in Appendix D in the book is dedicated to Peirce [Ogden and Richards 1946, pp279–290].
28. As discussed in one of his biographies [Marchand 1989, pp38–39]. This was at Trinity Hall college.
29. [Sebeok 1972, p13].
30. Morris presented this in his 1938 book *Foundations of the Theory of Signs*, which is reprinted in *Writings on the General Theory of Signs* [Morris 1971].
31. The 1946 book is reprinted as part of *Writings on the General Theory of Signs* [Morris 1971, pp74–400]. The quote is on page 301.
32. [Hodge and Kress 1988, p14].
33. Perhaps due to the continuous and rapid developments of ideas in the field of semiotics during the mid-20th century, the books of Morris and Sebeok are often bundled articles, mostly unaltered but usually with a reflective intro-duction and/or afterword, and in other cases revised and edited. While it is interesting to see how insights and ideas developed (I have taken care to date the sources appropriately), it is sometimes a bit harder to find the coherence in the material.
    The quote was in an article 'Animal Communication – a communication net-work model for language is applied to signalling behavior in animals' [Sebeok 1965] and reprinted as a chapter in the book *Perspectives in Zoosemiotics* [Sebeok 1972, pp63–83].
34. [Sebeok, Hayes, and Bateson (eds.) 1964, pp275–276].
35. [Sebeok, Hayes, and Bateson (eds.) 1964, pp277–287].
36. *Perspectives in Zoosemiotics* [Sebeok 1972].
37. Sebeok's communication network framework [Sebeok 1965], [Sebeok 1972, pp67–70], is presented in semiotic literature [Eco 1976, p175], [Cobley and Jansz 1997, p128], sometimes even without mentioning the source in the case of Sean Hall's *This Means This, This Means That* [2012, p16].
38. Sebeok has published various versions of the article on 'digital and analog coding' in communication, I've used the chapter 'Coding in the Evolution of Signalling Behavior in Perspectives in Zoosemiotics' [Sebeok 1972, pp7–33].
39. Jakobson presented a paper 'Linguistics and Poetics' at a conference in 1958, later revised and published in *Style in Language* [Sebeok (ed.) 1960, pp350–377].
40. [Petrilli and Ponzio 2001, p28].
41. [Cobley and Jansz 1997, pp126–129].
42. [De Waal 2016, pp7–9], [Dennett 2013, pp70–71].
43. [Norris, 2004].

44. [Sebeok and Danesi 2000, p10].
45. [Lane (ed.) 1970].
46. [Eco 1976, p58].
47. [Hall 1983, pp59–63].
48. Umberto Eco in *a Theory of Semiotics* [1976, p56 and footnote on p143] discusses the different meanings of the word, and writes about Coding Contexts and Circumstances [1976, pp110–113].
49. The example is also in the HCI textbook of Dix et al. [1993, p104].
50. Nørretranders first introduced the term 'exformation' in *The User Illusion* [1998, p92], and developes it throughout the book.
51. In an article for *Forum* magazine 'The Medium is the Message' [McLuhan 1960] and later in *Understanding Media* [1964]. See also *The Mechanical Bride – folklore of industrial man* [McLuhan 1951], an analysis of the effects of the mass media on society in the 1950s.
52. [Rosengren 2000, p30].
53. [Eco 1976, p20].
54. [Eco 1976, pp115–121].
55. [Bolt, 1980].
56. Translated by Jo Koch from the Danish text from 1971 as *Life Between Buildings* [Gehl 1987, pp133–145].
57. Dutch photographer Jan Dirk van den Burg published a beautiful book with images of the most amazing shortcuts people take [2011].
58. [Barfield 1993, p230].
59. In a paper for the *19th International Symposium of Electronic Art (ISEA)*, which was held in Sydney in June 2013 [Bongers 2013].
60. See for instance *Multimodal Discourse, the modes and media of contemporary communication.* [Kress and van Leeuwen 2001].
61. [Jewitt (ed.) 2009, Ch. 5].
62. The social semiotic writings of Theo van Leeuwen reflect his background as a film maker, and a musician [1999, Ch. 6].
63. [Mulder 2010, 2013].
64. [McLuhan 1960], [McLuhan 1964, p7].
65. In *Understanding Media*, Chapter 2, 'Media Hot and Cold' [McLuhan 1064, p24].
66. [McLuhan and Fiore, 1967]. They also created a sonic version, "conceived and co-ordinated" by Jerome Agel (who also played a big role in conceiving the book), and produced and composed by John Simon, a very dense and fascinating 'assemblage' of voices, instruments, samples, and McLuhan reading his texts and aphorisms. It was recorded in April and May 1967. Biographer Philip Marchand calls it "one of McLuhan's more offbeat projects" and likened it to an equivalent of an (imaginary) McLuhan music video [Marchand 1989, p187]. It was originally released as an LP by Columbia Records in 1968, and in 2013 re-issued (on vinyl) with a booklet with essays of Paul D. Miller (DJ Spooky) among others, and photos of the recording sessions. The recording can also be found on the internet.
   Later attempts to bring writing and layout at the same level of semiotic expressions were less convincing, such as Counterblast in the late 1960s, designed by Harley Parker [McLuhan 1969] [Coupland 2009, p169], or as

a recent example Bruce Sterling's collaboration with designers in *Shaping Things* which sometimes obscures or influences the terrific writing [Sterling 2005]. There are other examples, where the layout and typography can actually obscure the content of the writing – the medium is the mirage, one could say. One of these examples is my own experience of my first book *Interaction with our Electronic Environment* [2004], which was published in a series of small books ("Cahier", see Chapter 1, in the section 'Interaction and Education'. Dutch graphic designers DieTwee ('those two') [Staal (ed.) 2002], known for their radical yet insightful approach to design, had recently developed a new style and template for the series which looked great but wasn't always doing justice to the content (including challenging conventions such as the use of italics, replacing these with a different font colour).

67. See for instance the journal papers 'The Projector as Instrument' [Bongers, 2011] and 'Interactive Video Projections as Augmented Environments' [Bongers 2012].

68. The picture of Ron Mueck's work in the *Sensation* catalogue shows just the image [Rosenthal (ed.) 1997, p127], and so does the catalogue of Mueck's exhibition at the National Gallery of Victoria in Melbourne, but it also shows the work in two different gallery views which gives some of the scale [Hurlston (ed.) 2010].

69. The book *Ways of Seeing* contains seven essays, three of which use only images, by John Berger, Sven Blomberg, Chris Fox, Michael Dibb, and Richard Hollis. It is based on the BBC television series with the same name [Berger et al. 1972].

70. A good overview of course is in *The Story of Art* of Ernst Gombrich [1995].

71. [Lasserre 2018, p10]. The notion of onomatopoeia is also covered in the *Soundscapes* book by Murray Schafer [1977, pp40–42].

72. [Eco 1976, p9].

73. [Hall 1966].

74. [Deutscher 2005, pp116–117].

75. [Deutscher 2005, pp115–143].

76. [Hodge and Kress 1988, p16].

77. In an article in *The Guardian* 'Why It's Time to Stop Worrying about the Decline of the English Language' [Shariatmadari 2019], adapted from Shariatmadari's new book *Don't Believe a Word: the surprising truth about language*. The article appeared later in *The Guardian Weekly* as 'Lingua Fracas!!', on 23 August 2019, pp34–39.

78. [Erickson 1990].

79. [Nelson 1990].

80. [Kay 1990, p200].

81. Alberto Alessi describes in *The Dream Factory* several of these items, in the section on Philippe Starck [Alessi 1998, pp74–79].

82. [Fukasawa (ed.) 2007, pp19–21].

83. [Gross et al. 2014].

84. [Hodge and Kress 1988, p22], [Eco 1976, p178].

85. [Hodge and Kress 1988, p22], Gunther Kress in *Multimodality – a social semiotic approach to contemporary communication* [Kress 2010, pp62–69], see also Kress's paper 'Against Arbitrariness' [1993]. Kress passed away in June 2019.

86. [Eco 1976, p178].
87. [Morris 1971 p273] (in the original 1946 version, 7.2), [Eco 1976, p192].
88. [Eco 1976, p200], Eco refers to J. J. Gibson's second book [1966].
89. In *Kant and the Platypus - essays on language and cognition* [Eco 1999, p161, p407].
90. [Eco 1976, p180].
91. [Eco 1999, pp276–279].
92. [Eco 1976, p218].
93. [Dix et al. 2004], [Jacko (ed.) 2012].
94. [Sharp, Rogers, and Preece 2007, p673].
95. [Shaer and Honecker 2009, pp69–71].
96. [de Souza et al. 2001], [de Souza 2005], [de Souza et al. 2009], see also [Valtolina et al. 2012].
97. [Scolari 2001, 2009].
98. [Kay 1990].
99. This was said during a panel discussion at the *TEI* (Tangible and Embedded Interaction) conference in 2011, and see also his ID Sketchbook [2009, p21], but it goes back to his lecture notes from a course he taught with Terry Winograd at Stanford University in 1992 (which Bill kindly shared with me), though at that time using the term 'gestural' instead of 'tangible'.
100. [Kress and van Leeuwen 2001, 2006], [Norris 2004], [Kress 2010], [Jewitt 2009], [Bateman 2008].
101. [Bernsen 1993], [Schomaker et al. (eds.) 1995].
102. [Hodge and Kress, 1988, p264].
103. [Kress and van Leeuwen 2006, p154], [van Leeuwen 1999, Chapter 7, p156].
104. [Hodge and Kress, 1988, p26, p27].
105. [Eco 1976, p7].
106. [Eco 1976, pp58–59, pp178–179].
107. In *Understanding Media* [McLuhan 1964, pp142–143].
108. See Banksy's book *Wall and Piece* [Banksy 2006, pp116–117].
109. [Banksy 2006, p184], [Ellsworth-Jones 2012, p18].
110. [Eco 1976, p179].
111. [Eco 1976, p178].
112. [Eco 1976, p181], see also Walter Benjamin [2008].
113. In *The Other Side of Painting – Marcel Duchamp* [García-Bermejo 1996, p30], and in Banksy's *Wall and Piece* [2006, p27].
114. Benjamin wrote the essay in 1935 and revised it in 1939. I used a recent translation published by Penguin [Benjamin 2008].
115. [García-Bermejo 1996, p14, p21].
116. The unicycle in the video was 'wheeled' with great skill by Simon O'Brien, and also presented an AR (augmented reality) version made for Shaun Gladwell's show *Pacific Undertow* in the Museum of Contemporary Art in Sydney in 2019. By looking through my smartphone (or tablet), as a visitor I could see O'Brien ride the unicycle in the actual space where the sculpture was placed. The installation is discussed in an essay in the catalogue of the exhibition [Thwaites 2019, p150].
117. [Rosenthal (ed.) 1997].
118. The image is based on a photograph of an auction in 1987 at Christie's selling for the then record price of over £22m one of Vincent van Gogh's sunflower still-life paintings from 1889 [Ellsworth-Jones 2012, p186].

119. [de Goede (ed.) 2007, pp38–53].
120. This was for the Biennale in Sydney in 2016, the act took place outside the Carriage Works theatre. Although I missed the performance (only saw it on video), later in the week I collected some left-overs snippets if anybody wants to make a bid … but that would defeat the point.
121. [Potter 2019, p245].
122. [Hodge and Kress 1988, p20], [Eco 1976, p24].
123. Daniel Dennett in *Consciousness Explained* [1991, p234].
124. [Deutscher 2010, pp129–156].
125. In his book *Language – the cultural tool* [Everett 2012] and also in the documentary *The Amazon Code* for the Smithsonian Channel from 2012 (which can be found online but it is a bit hard to search for as most hits will relate to the mail-order universe of the company Amazon).
126. [Deutscher 2010, p143].
127. J. J. Gibson is critical about this too, in *The Ecological Approach to Visual Perception* he writes: "Time is not another dimension of space, a fourth dimension, as modern physics assumes for reasons of mathematical convenience." [1979, p101].
128. [Hall 1983, pp81–90].
129. [Kelly 2016, p54, p260].
130. Guy Deutscher in *Through the Language Glass* [2010, p153].
131. [Deutscher 2010, p155].
132. [Dixon 2019, p113].
133. The emergence of colour names, and the reasons for it (anatomical, cultural, but not hierarchical as once thought), are discussed by Deutscher in chapter 4, 'Those Who Said Our Things Before Us' [2010, pp79–98].
134. A good overview is given by Stefan Käufer and Anthony Chemero in *Phenomenology – an introduction.* [2015].
135. [Deutscher 2010, pp226–229].
136. [Deutscher 2010, pp229–230].
137. [Deutscher 2010, pp222–226], [Winawer et al. 2007]. There is a great TED talk by Lera Boroditsky, the leader of this research, for TEDWomen in November 2017.
138. [Donker et al. 2015].
139. [Haviland 1998].
140. [Deutscher 2010, p160], [Dixon 2019, p145].
141. [Lawlor 1991, p105].
142. [Deutscher 2010, p172].
143. [Lawlor 1991, p268], [Dixon 2019].
144. The documentary *The Balanda and the Bark Canoes* from 2006 by Molly Reynolds, Tania Nehme, and Rolf de Heer. The quote by 'Rolf de Ralf' (as he is being called) is at 0:34 minutes. 'Balanda' is a word in the Yolngu language, meaning 'stranger', which reveals ancient trading connections between the peoples in the north of the Australian continent and Indonesian people, as it is borrowed from a Malay word. In fact, in Bahasa Indonesia, 'Belanda' means Dutch, from 'Hollander' (English is 'Inggeris', while a generic foreigner is 'asing').
145. [Dixon 2019, p64].

146. [Lawlor 1991, p244].

147. [Dixon 2019, pp61–71].

148. This notion was introduced by Eco in *Semiotics and the Philosophy of Language*, in Chapter 2, 'Dictionary vs. Encyclopedia' [1984, pp46–86].

149. [Eco, 1999, p89, pp92–93, pp241–251], Eco draws from the work of Steven Jay Gould [1991, Chapter 18].

150. Uluru is the original name in the Pitjantjatjara language, and has very strong cultural significance [Lawlor 1991, p160]. The British settlers named it Ayers Rock when they first sighted it in 1873, and it has been known as such for a long time, until 1993 when the original name was restored. It now officially has both names, but of course Uluru is preferred. Eco uses the English name, which makes his point of the mountain being a rock stronger [Eco, 1999, pp224–225]. Uluru was handed back to the Pitjantjatjara people in 1985 [AIATSIS and Pascoe 2018, p159].

151. Malinowski's essay [Ogden and Richards 1946, pp297–336] (the section on phatic communication starts on page 315), is entitled 'The Problem of Meaning in Primitive Language', and this sort of choice of words including 'savages' nowadays sound like an affront. But as discussed in Chapters 3 and 4, Darwin's ideas on evolution were hijacked to make false claims and at the time it was very common to think in this hierarchical way. It was considered that the classical languages such as ancient Greek and Latin were superior, and even modern European languages as degenerates of this [Deutscher 2010, p132]. We now know that this is completely misguided, and as several linguists and anthropologists have shown, there is nothing primitive about the language of Indigenous people [Deutscher 2005, 2010], [Everett 2012], [Dixon 2019]).

152. As ethologist Frans de Waal describes in his books particularly on social behaviour of primates [De Waal 2016, 2019].

153. [Linden 2015, pp16–19]. This is also discussed in Chapter 9 *Experiencing* in the section on Tactual Perception, and in Chapter 11, *Expressing*.

154. [Aiello and Dunbar 1993], [Mithen 2005, pp134–136].

155. [Everett 2012, pp236–239].

156. *Small Talk Survival – praktische gids voor de gesprekken tussendoor* (practical guide for in-between conversations) by Liz Luyben and Iris Posthouwer [2019].

157. [Eco 1976, p60].

158. [Gibson 1966, p285], [Gibson 1950], [Gibson and Crooks 1953], [Gibson 1977, 1979, Ch. 8].

159. [Norman 1988, 2013], further references are in Chapter 9 *Experiencing*, in the section *Affordances in HCI*.

160. See for instance Daniel Kahneman's *Thinking, Fast and Slow* [2011, Malcolm Gladwell's *Blink* [2005], Leonard Mlodinow's *Subliminal* [2012], and Daniel Goleman's *Emotional Intelligence* [1995] and *Focus* [2013]. This is explored in depth in Chapter 10 *Thinking*.

161. [Eco 1976, pp177–178].

162. In *The Ecological Approach to Visual Perception* [Gibson 1979, p128].

163. [Eco 1999, p161].

Interface

Connecting Volume 1 and Volume 2— Preliminary Frameworks and Directions

Imagine a person who gets marooned on a desert island, without any tools other than a smartphone, which will be useless from the moment the battery runs out. All the person has access to are materials to make tools, from stone, wood, fibres, maybe bone (preferably from another animal). Maybe by using fire, and finding iron ore, metal tools can be made. (This seemed so much easier in *Minecraft*...) Then maybe mechanical contraptions, machines perhaps, and using the machines to make other machines. It took humankind millions of years to get to that point, but with the knowledge we have now, a replay can be a bit quicker. It will however take a lot longer than an individual lifetime to develop electrical technology to power the smartphone. Many more people are needed, technical developments are inseparable from culture, from social development.*

In the previous chapters the different stages of human development have been discussed. They are not independent, successive stages as we have seen – the key factors (tool use, language development, culture) all started millions of years ago. The technocultural periods that were identified, and reflected on in Chapters 3–6, are about the different way we related to our technological and natural environments. We first co-evolved with our tools and our communication as hunter-gatherers, then settled and lived off agriculture, followed by the industrialisation period. In the current period, most of us live in cities, and spend a lot of our time indoors, often sitting down. As Vybarr Cregan-Reid points out in his recent book

---

* Many action hero and science fiction movies show the unrealistic view of the lone inventor, creating a whole new technology entirely by himself (usually a 'he', yes), sometimes with the help of the robots and machines he makes, and maybe a butler or a sidekick. Examples are *Iron Man* (a metal suit that flies, weapons), *Batman Begins* (protective clothes, weapons, vehicles), *I am Legend* (a serum to cure zombies), *Surrogates* (cloning the family's minds and putting them in robot copies), *Transcendence* (uploading the mind in a computer and then using robots to build robots and a new body), and of course Melvin in *Captain Underpants*. In the real world, any level of technological advancement needs teamwork and multiple parties to support it, and a lot of time.

Primate Change: How the World We Made Is Remaking Us[1] and as discussed in the earlier chapters in this book, we spent the longest time in the first period. In this period, we co-evolved with our environment, and have been using the same body and mind in the completely different environments that we've created. We first lived in nature, then worked in the fields, then worked in factories, and now work in offices. As Cregan-Reid puts it, physical activity diminished at every step, which he calls "lifestyle revolutions": the hunter ("human version 1.0"), the farmer ("agricultural body", "human 2.0"), the worker ("industrial body", "human 3.0"), and the office worker ("human 4.0"). Luckily we invented sports, fitness, jogging, and activity trackers to measure it all. We also still do what we are good at: making tools, and being social – we have evolved to communicate, though our voice, gestures, but also our dynamic eyebrows, exposed eye-whites, facial expression, posture, and touch. While our tools have always extended our ability to influence our environment, our media have extended our communication.

In this first part of *Understanding Interaction*, Volume 1, we've explored where we have come from, and how we arrived at where we are now. With this knowledge, in Volume 2, we explore who we are and where we are going – taking a close look at how we physically and mentally interact with our environment. This means looking into human psychology, biology, physiology, and neurology (Chapter 7 *Being*), how we sense the world (Chapter 8 *Feeling*), how we perceive (Chapter 9 *Experiencing*), how we process information and reason (Chapter 10 *Thinking*), and how we influence our environment and other people (Chapter 11 *Expressing*).

The final chapter (12 *Understanding*) is concerned with design directions, frameworks, and future directions. This is where the full interaction framework is presented. It is an extension of my 'multimodal interaction space', which describes any (part of an) interaction in modes, modalities, and levels.[2] The **modes** have been covered in the last chapter, Chapter 6 *Communicating* as the (re-)presentation spectrum: from the actual object and its affordances, to representations from mimetic (iconic) to abstract (symbolic). The **modalities**, based on our senses (many more than the proverbial five) and our expressive abilities, are covered in Volume 2 (Chapters 8 *Feeling*, 9 *Experiencing*, and 11 *Expressing*). The **levels** are the steps (sequential or conceptual) that take place in an interaction, from the needs and goals formulated to concrete actions, which is supported by Chapter 10 *Thinking*.

A further expansion of the simplified 'interaction loop' as shown in the diagram in Figure I.1 is what I call the '**reach** of interaction'. While the modes and modalities are about the width of the interaction ('interaction bandwidth'), the levels are about the height, and the 'reach' is about the depth of the interaction. When interacting with a basic system, such as a switch turning on the light, there is a direct reaction (rather than inter-action). A more advanced system can have its own behaviours, processes, decision making, etc., and the response is less predictable (but hopefully more appropriate). In this case, we 'reach' further into the system.

The purpose of exploring all the possibilities in this interaction space is to create a design space, using this expanded set of rich palettes (of different modes, modalities, levels, and reach). The aim of optimising and even expanding interactions is supported by having these palettes at our disposal.

We can also interact or communicate *through* a system, which acts as a medium, 'reaching' to another person or audiences. Examples are the telephone, a video meeting or broadcast, or a musical instrument.[3] At the beginning of the COVID-19 pandemic in early 2020, we were all forced to use video connections as our main mode of communicating, and it was interesting to see, in spite of all its obvious strengths and virtues, how video-communication software severely limited the bandwidth of interaction. Due to its focus on face-to-face communication, we were all missing the 'elephant in the Zoom'. This led to experiments with multiple video streams and large screens, microphone placement and better speakers, to create an increased sense of presence between the locations.

Media can reach out to us too, as they do in traditional broadcasts (radio, TV) and recently in internet-based media such as the World Wide Web, blogs, podcasts, and social media. Here the reach is inverted, as it were – as mass media try to reach the receiver (as in McLuhan's aphorism *The Medium is the Message*), we might feel little control and opportunity for participation. McLuhan conflated this level of participation with density (interaction bandwidth through modes and modalities) in his definition of 'hot' and 'cool' media.[4]

The opportunities of the 'reach' deep into a system are most strongly shown in the case of AI, artificial intelligence, where the potential for interaction is greatest. It is also the case with the greatest *need* for interaction, as otherwise the AI will just do its own thing, often in all its ignorance; this would be a missed opportunity. Chapter 12 *Understanding* also explores the notion of transhumanism, where humans and technology merge as has been pioneered by the notion of cyborgs since the 1960s.[5] By accepting and embracing the vast complexity of the human body and mind, all the physiological and neurological processes, a more symbiotic relationship can be established. This is in contrast to the attitudes of the scientific revolution, Enlightenment, and industrialisation, as discussed in Chapter 5 in the section Humans versus Nature, leading to a tendency to disrespect nature. What is being recreated is this reduced and dumbed-down version of nature ("denuded" as Lewis Mumford already called it in the 1930s),[6] and if this leads to a version of posthumanism (the world inhabited by artificial creatures that we invent) that would be a future even more horrible than depicted in science fiction. I've called this the 'Frankenstein Scenario', where technology takes over the world while not being ready for it.[7] Recent publications have been more critical about transhumanism and AI, such as Mark O'Connell's *To Be a Machine* and Harry Collins's *Artifictional Intelligence*, while thinkers such as Ray Kurzweil have been actively predicting the 'singularity' where computers become more intelligent than humans for several decades, but also Yuval Noah Harari in *Homo Deus* and James Lovelock in *Novocene* are optimistic about this.[8] But it seems that this version of singularity is mostly based on the misguided and limited view of human capabilities – and nature. Chapter 12, in Volume 2, reflects on these scenarios in detail, from the historical perspective as presented in the chapters of Volume 1, to the in-depth exploration of human biology, physiology, neurology, and psychology in the chapters of Volume 2.

NOTES

1. [Cregan-Reid 2018].
2. [Bongers and van der Veer 2007].
3. [Bongers 2000].
4. As discussed in Chapter 6 *Communicating* in the section Semiotic Resources and Framing; see *Understanding Media*, Chapter 2, 'Media Hot and Cold' [McLuhan 1964, p24].
5. For instance the pioneering work of Australian performance artist Stelarc, who extended his body with various robotic technologies, such as a third arm. We met a couple of times, he is a very generous person with fascinating ideas that really crossed boundaries – largely his own. When he states "the body is obsolete", I would normally disagree but in his case I appreciate it because he experiments on his own body. Where my earlier work with musical instruments such as gloves, hands, and other wearables – with which we tried to get as close to the performer's body as possible – always stayed outside the human body, Stelarc went a step further in connecting to technology at a deeper level, under the skin, inside his body. We invited Stelarc to do a keynote presentation at the *NIME++* (*New Interfaces for Musical Expression*) conference in 2010 in Sydney [Beilhartz et al. (eds.) 2010]. A good overview of his earlier work, and reflections on the cyborg nature, is in *Stelarc the monograph*, edited by Marquard Smith [2005].
6. In *Technics and Civilisation* [Mumford 1934].
7. "To invent its own successor may be seen as the ultimate goal of a species, but to be taken over by technology prematurely is a Frankenstein scenario." In *Interaction with our Electronic Environment – an ecological approach to physical interface design* [Bongers 2004, p6]. Computer viruses can be seen as an early example of that, as Tor Nørretranders has already noted in the 1990s they are an example of Artificial Life [1998, p361], and as such they are surprisingly successful.

8. *To Be a Machine – adventures among cyborgs, utopians, hackers, and the futurists solving the modest problem of death* [O'Connell 2016], *Artifictional Intelligence – against humanity's surrender to computers* [Collins 2018], *The Age of Spiritual Machines – when computers exceed human intelligence* [Kurzweil 1999] (thanks to former MDes student Laura Hindes for giving me this book in 2014), *Homo Deus – a brief history of tomorrow* [Harari 2016], *Novacene – the coming age of hyperintelligence* [Lovelock 2019].

References

Aarts, E. H. L., Harwig, H. A., and Schuurmans, M. F. H., Ambient intelligence. In: Denning, P. J. (ed.), *The Invisible Future - the seamless integration of technology in everyday life*, pp236–250, McGraw-Hill, 2002.

Aarts, E. H. L., Collier, R., van Loenen, E., and de Ruyter, B. (eds.), Ambient intelligence. In: *Proceedings of the First European Symposium*, Springer Verlag, 2003.

Aarts, E. H. L., and Marzano, S. (eds.), *The New Everyday - views on ambient intelligence*. NAI 010 Publishers, 2003.

Aarts, E. H. L., and Diederiks, E., *Ambient Lifestyle - from concept to experience*. BIS Publishers, 2006.

Ackerman, R. R., Mackay, A., and Arnold, M. L., The hybrid origin of "modern" humans. *Evolutionary Biology*, 43/1, pp1–11, March 2016.

Addlesee, M., Curwen, R. Hodges, S., Newman, J., Steggles, P., Ward A., and Hopper, A., Implementing a sentient computing system. *IEEE Computer*, 34/8, pp2–8, August 2001.

AIATSIS and Pascoe, B., *The Little Red Yellow Black Book - an introduction to indigenous Australia*. 4th edition, Australian Institute for Aboriginal and Torres Strait Islanders, Aboriginal Studies Press, 2018.

Aiello, L. C., and Dunbar, R. I. M., Neocortex size, group size and the evolution of language. *Current Anthropology*, 34/2, pp184–193, April 1993.

Alessi, A., *The Dream Factory – Alessi since 1921*. Könemann, 1998.

Anich, P. S., Anthony, S., Carlson, M., Gunnelson, A., Kohler, A. M., Martin, J. G., and Olson, E. R., Biofluorescence in the platypus. *Mammalia*, 85/2, pp179–181, 15 October 2020.

Attiwill, P., and Wilson, B. (eds.), *Ecology, an Australian Perspective*. Oxford University Press, 2003.

Aukstakalnis, S., and Blatner, D., *Silicon Mirage, the Art and Science of Virtual Reality*, Peachpit Press, 1992.

Baecker, R. M., and Buxton, W. A. S., A historical and intellectual perspective. In: Baecker, R. M., and Buxton, W. A. S. (eds.), *Readings in Human-Computer Interaction*, pp41–54, Morgan Kaufman, 1987.

Baecker, R. M., and Buxton, W. A. S., Case study D: The Star, the Lisa, and the Macintosh. In: Baecker, R. M., and Buxton, W. A. S. (eds.), *Readings in Human-Computer Interaction*, pp649–652, Morgan Kaufman, 1987.

Bahn, P., *Archaeology, a very short introduction.* Oxford University Press, 2012.

Banksy, *Wall and Piece.* Century, Random House, 2006.

Banzi, M., and Shiloh, M., *Make: Getting Started with Arduino - the open source electronics prototyping platform.* 3rd edition, Maker Media Inc., 2014.

Barfield, L., *The User Interface – concepts and design.* Addison Wesley, 1993.

Barfield, W., and Danas, E., Comments on the use of olfactory displays for virtual environments. *Presence,* 5/1, pp109–121, 1996.

Bateman, J. A., *Multimodality and Genre - a foundation for the systematic analysis of multimodal documents.* Palgrave McMillan, 2008.

Bauby, J.-D., *The Diving Bell and the Butterfly – a memoir of life in death.* Translated from the French, *Le Scaphandre et le Papillion* by Jeremy Leggat. Vintage Books, 1997.

Baumrind, D., Some thoughts on ethics of research - After reading Milgram's "behavioral study of obedience". *American Psychologist,* 19/6, pp421–423, June 1964.

Beilharz, K., Johnston, A., Ferguson, S., and Chen, Y.-C. (eds.), *Proceedings of the Conference on New Interfaces for Musical Expression,* NIME++, UTS, June 2010.

Benjamin, W. B. S., *The Work of Art in the Age of Mechanical Reproduction,* translated from the German Das Kunstwerk im Zeitalter seiner technischen Reproduzierbarkeit 1939 by J. A. Underwood, Penguin, 2008.

Berger, J., Blomberg, S., Fox, C., Dibb, M., and Hollis, R., *Ways of Seeing.* BBC and Penguin, 1972.

Berger, L. R., Hawks, J., de Ruiter, D. J., Churchill, S. E., Schmid. P., Delezene, L. K., Kivell, T. L., Garvin, H. M., Williams, S. A., DeSilva, J. M., and Skinner, M. M., *Homo naledi,* a new species of the Genus *Homo* from the Dinaledi Chamber, South Africa. *eLife,* 10, September 2015.

Bernsen, N. Ø. Modality theory: Supporting multimodal interface design. In: *Proceedings from the ERCIM Workshop on Multimodal Human-Computer Interaction,* Nancy, pp13–23, November 1993.

Bickerstaff, C., *Interface: people, machines, design.* Museum of Applied Art and Sciences, 2014.

Blockley, D. *Engineering - a very short introduction.* Oxford University Press, 2012.

Boersema, J. J., *Beelden van Paaseiland – over de duurzaamheid en veerkracht van een cultuur.* Atlas Contact, 2020.

van Bogaert, P., *Feel: a manual.* Special issue of the AS Mediatijdschrift, 169, MuHKA Antwerp, 2004.

Bolanowski, S. J., Gescheider, G. A., Verrillo, R. T., and Checkosky, C. M., Four channels mediate the mechanical aspects of touch. *Journal of the Acoustical Society of America,* 84/5, pp1680–1694, November 1988.

Bolt, R. A. "Put-that-there": Voice and gesture at the graphics interface. In: *Proceedings of the 7th Annual Conference on Computer Graphics and Interactive Techniques SIGGGRAPH,* pp262–270, 1980.

Bongers, A. J., The use of active tactile and force feedback in timbre controlling musical instruments. In: *Proceedings of the International Computer Music Conference (ICMC),* Århus, Denmark, pp171–174, 1994.

Bongers, A. J., An interview with Sensorband. *Computer Music Journal,* 22/1, pp13–24, 1998.

Bongers, A. J., Tactual display of sound properties in electronic musical instruments. *Displays Journal*, 18, pp129–133, 1998.

Bongers, A. J., Eggen, J. H., Keyson, D. V., and Pauws, S. C., Multimodal interaction styles. *HCI Letters Journal*, 1/1, pp3–5, 1998.

Bongers, A. J., Interaction in multimedia art. *Knowledge-Based Systems Journal*, special issue on Creativity and Cognition. Elsevier Press, 13/7–8, pp479–485, December 2000.

Bongers, A. J., Physical interaction in the electronic arts, interaction theory and interfacing techniques for real-time performance. In: Wanderley, M. M., and Battier, M., (eds.), *Trends in Gestural Control of Music*. IRCAM Paris, pp41–70, 2000.

Bongers, A. J., Interactivating spaces. In: *Proceedings of the Systems Research in the Arts Conference*, Baden-Baden, Germany, August 2002.

Bongers, A. J., and Harris, Y. C., A structured instrument design approach: The video-organ. In: *Proceedings of the Conference on New Instruments for Musical Expression (NIME)*, Dublin, Ireland, May 2002.

Bongers A. J., Interactivating spaces/strategie interattive, *L'Arca Edizione*, *International Magazine of Architecture, Design and Visual Communication*, June 2003.

Bongers A. J., *Interaction with our Electronic Environment – an e-cological approach to physical interface design*, Cahier Book Series No 34, Department of Journalism and Communication, Hogeschool van Utrecht, March 2004.

Bongers, A. J., Palpable pixels, a method for the development of virtual textures. In: Ballesteros, S., and Heller, M. A. (eds.),*Touch, Blindness and Neuroscience*, UNED Ediciones, 2004.

Bongers, A. J., *Interactivation - towards an e-cology of people, our technological environment, and the arts*. Ph.D. Thesis, Vrije Universiteit Amsterdam, July 2006.

Bongers, A. J., Electronic musical instruments: Experiences of a new luthier, *Leonardo Music Journal*, 17, pp9–16, 2007.

Bongers, A. J., and van der Veer, G. C., Towards a multimodal interaction space - Categorisation and applications. Special Issue on Movement-Based Interaction of the *Journal of Personal and Ubiquitous Computing*, 11/8, pp609–619, 2007.

Bongers, A. J., and van der Veer, G. C., HCI and design research education - A creative approach. In: Kotzé, P. (ed.), *Creativity and HCI: From Experience to Design in Education*, pp91–105, Springer Verlag, 2009.

Bongers, A. J., The projector as instrument. *Journal of Personal and Ubiquitous Computing*, Special Issue on Personal Projection, 16/1, pp65–75, 2011.

Bongers, A. J., and Mery-Keitel, A. S., Interactive kaleidoscope: Audience participation study. In: *OzCHI Conference Proceedings*, Canberra, pp58–61, December 2011.

Bongers, A. J., and Smith, S. T., Interactivating rehabilitation through active multimodal feedback and guidance. In: Röcker, C., and Ziefle, M. (eds.), *Smart Healthcare Applications and Services - developments and practices*, Chapter 11, pp236–260, IGI Global, 2011.

Bongers, A. J., Interactive video projections as augmented environments. *International Journal of Arts and Technology*, 15/1, pp17–52, 2012.

Bongers, A. J., Anthropomorphic resonances – On the relationship between computer interfaces and the human form and motion. *Interacting with Computers Journal*, special issue on Organic User Interfaces, Oxford University Press, 25/2, pp117–132, 2013.

Bongers, A. J., Traces - Reading the environment. In: Cleland, K., Fisher, L., and Harley, R. (eds.), *Proceedings of the 19th International Symposium of Electronic Art*, Sydney, June 2013.

Bongers, A. J., Smith, S. T., Donker, V., and Pickrell, M., Interactive rehabilitation tiles. In: *Work in Progress paper, Proceedings of the 8th International Conference on Tangible, Embedded and Embodied Interaction (TEI)*, ACM, München, February 2014.

Bongers, A. J., Smith, S. T., Donker, V., Pickrell, M., Hall, R., and Lie, S., Interactive infrastructures – Physical rehabilitation modules for pervasive healthcare technology. In: Holzinger, A., Ziefle, M., and Röcker, C. (eds.), *Pervasive Health – state of the art and beyond*, pp229–254, Springer, 2014.

Bongers, A. J., and Heffer, C., Pattern Stations - Extending textile materials through tangible interaction. In: *Proceedings of the 9th International Conference on Tangible, Embedded, and Embodied Interaction*, Stanford, CA, USA, pp405–406, January 2015.

Bongers, A. J., Tangible landscapes and abstract narratives. In: *Proceedings of the of the 14th International Conference on Tangible, Embedded, and Embodied Interaction (TEI)*, pp689–695, 2020.

Bonta, M., Gosford, R., Eussen D., Fergeson, N., Loveless, E., and Witwer, M., Intentional fire-spreading by "Firehawk" raptors in Northern Australia. *Journal of Ethnobiology*, 37/4, pp700–718, December 2017.

Bowman, D., and Davies, J., Indigenous land use. In: Attiwill, P., and Wilson, B. (eds.), *Ecology, an Australian Perspective*. Chapter 25, pp406–420, Oxford University Press, 2003.

Bramble, D. M., and Lieberman, D. E., Endurance running and the evolution of *Homo. Nature* 432, 18 November, pp345–352, 2004.

Bregman, R. C., *De Geschiedenis van de Vooruitgang*. Bezige Bij, 2013.

Bregman, R. C., *De Meeste Mensen Deugen - een nieuwe geschiedenis van de mens*. De Correspondent, 2019.

Bregman, R. C., *Gratis Geld voor Iedereen – hoe utopische ideeën de wereld veranderen*. 1st edition, 2014, De Correspondent, 2020.

Bregman, R. C., *Humankind - a hopeful history*. Translation of *De Meeste Mensen Deugen* [2019], by E. Manton and E. Moore, Bloomsbury, 2020.

Brunt, T., Cecil, M., Griffiths, J., Adams-Hosking, C., and Murray, P., Where are the platypuses (*Ornithorhynchus Anatinus*) now? A snapshot in time of their distribution in the greater brisbane region. In: *Australian Mammology*, 43/3, pp368–372, 7 January 2021.

Bryson, W., *The Body - a guide for occupants*. Penguin, 2019.

Bryson, W., *A Short History of Nearly Everything*. Black Swan, 2003.

van den Burg, J. D., *Olifantenpaadjes - desire lines*. 2011.

Bush, V., As we may think. *Atlantic Monthly*, 176/1, pp101–108, July 1945.

Buxton, W. A. S., *Sketching User Experiences - getting the design right and the right design*. Morton Kaufmann, 2007.

Cane, S. B., *First Footprints - the epic story of the first Australians*. Allen & Unwin, 2013.

Carey, N., *The Epigenetics Revolution – how modern biology is rewriting our understanding of genetics, disease and inheritance*. Icon Books, 2011.

Carroll, J. M., Human computer interaction - Brief intro. In: Soegaard, M., and Dam, R. F. (Eds), *The Encyclopedia of Human-Computer Interaction*, 2nd edition, Chapter 2. Interaction Design Foundation, 2014.

Caruana, W., *Aboriginal Art*, New edition. Thames and Hudson, 2003.

Chadabe, J., *Electric Sound - the past and promise of electronic music*. Prentice-Hall, 1997.

Chatwin, B., *The Songlines*. Vintage Classics, 1987.

Clarke, P. A., *Where the Ancestors Walked - Australia as an Aboriginal landscape*. Allen & Unwin, 2003.

Cobley, P., and Jansz, L., *Introducing Semiotics*. Totem Books, 1997.

Colapinto, J., The interpreter - Has a remote amazonian tribe upended our understanding of language? *The New Yorker*, 16 April 2007.

Collins, H. M., *Artifictional Intelligence - against humanity's surrender to computers*. Polity, 2018.

Collins, N., *Handmade Electronic Music - the art of hardware hacking*. 2nd edition, Routledge, 2009.

Collins, N., Live electronic music. In: Collins, N., and d'Escriván, J. (eds.), *The Cambridge Companion to Electronic Music*. Chapter 4. 2nd edition, pp40–57, Cambridge University Press, 2017.

Cook, V. J., *Chomsky's Universal Grammar – an introduction*. Applied Language Studies, Blackwell, 1988.

Cooper, A., *About Face - The essentials of user interface design*. 1st edition, IDG Books, 1995.

Cooper, A., *The Inmates are Running the Asylum - why high-tech products drive us crazy and how to restore the sanity*. Sams Publishing, 1999.

Cooper, A., Reimann, R., Cronin, D., and Noessel, C., *About Face - the essentials of interaction design*. 4th edition, Wiley & Sons, 2014.

Coupland, D., *Marshall McLuhan*. Penguin, 2009.

Cregan-Reid, V. *Primate Change – how the world we made is remaking us*. Hachette, 2018.

Crutzen, P. J., Geology of mankind. *Nature* 415, p23, 2002.

Cumming, E., and Kaplan, W., *The Arts and Crafts Movement*. Thames & Hudson World of Art Series, first published 1991, revised edition 2002.

Darwin, C. R., *On the Origin of Species by Means of Natural Selection – or the preservation of favoured races in the struggle for life*. Penguin, 1968, first edition published 1859.

Davidson, I., and Noble, W., Tools and language in human evolution. In: Gibson, K. R., and Ingold, T. (eds.), *Tools, Language and Cognition in Human Evolution*, ch16, pp363–388, Cambridge University Press, 1993.

Davidson, R., *Tracks*. Book Club Associates, 1980.

Davidson, R., *Desert Places*. Viking, 1996.

Davidson, R., No fixed address – Nomads and the fate of the planet. *Quarterly Essay*, 24, pp1–53, 2006.

Davis, W., *The Wayfinders – why ancient wisdom matters in the modern world*. UWA Publishing, 2009.

Davison, N., The anthropocene epoch - Have we entered a new phase of planetary history? *The Guardian*, 30 May 2019.

Deleuze, G., and Guattari, F., *A Thousand Plateaus*. Translated from *Mille Plateaux* (1980) by Massumi, B., Bloomsbury Revelations edition (2013), 1988.

Dennett, D. C., *Consciousness Explained*. Little Brown, 1991.

Dennett, D. C., *Intuition Pumps and other tools for thinking*. Penguin, 2013.

Denning, P. J. (ed.), *The Invisible Future, the Seamless Integration of Technology Into Everyday Life*. McGraw-Hill, 2002.

Détroit, F., Mijares, A.S., Corny, J., Daver, G., Zanolli, C., Dizon, E., Robles, E., Grün, R., and Piper, P. J., A new species of *Homo* from the late pleistocene of the Philippines. *Nature*, 568, pp181–186, 2019.

Deutscher, G., *The Unfolding of Language – the evolution of mankind's greatest invention*. Arrow Books, 2005.

Deutscher, G., *Through the Language Glass - why the worlds looks different in other languages*. Arrow Books, 2010.

Diamond, J., *Guns, Germs and Steel – a short history of everybody for the last 13,000 years*. Vintage, 1997.

Diamond, J., *Collapse – how societies choose to fail or survive*. Penguin, 2005.

Diamond, J., *The World Until Yesterday – what can we learn from traditional societies?* Allen Lane, 2012.

Dirks, P. H. G. M., Roberts, E. M., Hilbert-Wolf, H., Kramers, J. D., Hawks, J., Dosseto, A., Duval, M., Elliott, M., Evans, M., Grün, R., and Hellstrom, J., The age of *Homo naledi* and associated sediments in the Rising Star Cave, South Africa. *eLife*, 6, 9 May 2017.

Dix, A., Finlay, J., Abowd, G., and Beale, R., *Human-Computer Interaction*. Prentice Hall, 1st edition. 1993, Pearson Education, 2nd edition. 1998, 3rd edition. 2004.

Dixon, R. M. W., *Australia's Original Languages - an introduction*. Allen & Unwin, 2019.

Donker, V., Markopoulos, P., and Bongers, A. J., REHAP balance tiles: A modular system supporting balance rehabilitation. In: *Proceedings of the 9th International Conference on Pervasive Computing Technologies for Healthcare*, Istanbul, Turkey, May 20–23, 2015.

Dorst, C. H., *Understanding Design - 175 reflections on being a designer*. BIS Publishers, 1st edition. 2003, 2nd edition. 2006.

Dorst, C. H., *Frame Innovation - create new thinking by design*. MIT Press, 2015.

Dourish, P., *Where the Action is - the foundations of embodied interaction*. MIT Press, 2001.

Duchamp, M., Apropos of 'Readymades', transcript of a talk given at a symposium in conjunction with the exhibition. *The Art and the Assemblage* at the Museum of Modern Art, New York, 19 October 1961. In: Affron, M. (ed.), *The Essential Duchamp*. Philadelphia Museum of Art/Yale University Press, 2018.

Dyson, F. J., *Disturbing the Universe*. Basic Books, 1979.

Eco, U., *A Theory of Semiotics*. Indiana University Press, 1976.

Eco, U., *Semiotics and the Philosophy of Language*, Macmillan Press, 1984.

Eco, U., *Kant and the Platypus - essays on language and cognition*. Random House, 1999.

Eggen, J. H. (ed.), *Ambient Intelligence in HomeLab*. Philips Research, 2002.

Eisenstein, E. L., *The Printing Revolution in Early Modern Europe*. Cambridge University Press, 1983.

Eisenstein, E. L., *Divine Art, Infernal Machine – the reception of printing in the West from first impressions to the sense of an ending*. University of Pennsylvania Press, 2011.

Ellsworth-Jones, W., *Banksy - the man behind the wall*. Aurum Press, 2012.

Enders, G. *Gut – the inside story of our body's most under-rated organ*. (translated from German *Darme mit Charme*, 2014), Scribe, 2015.

Engelbart, D. C., and English, W. K. A research center for augmenting human intellect. In: *Proceedings of the AFIPS Fall Joint Computer Conference*, San Francisco, CA, December 1998, pp395–410, 1969.

English, W. K., Engelbart, D. C., and Huddart, B., *Computer-aided Display Control*. Final Report, contract NAS 1-3988 SRI Project 5061, Stanford Research Institute Menlo Park California, July 1965.

Erickson, T. D., Working with interface metaphors. In: Laurel, B. (ed.), *The Art of Human-Computer Interface Design*, pp65–73, Addison-Wesley, 1990.

Everett, D. *Language – the cultural tool*. Profile Books, 2012.

Fitzmaurice, G. W., Ishii, H., and Buxton, W. A. S., Bricks: Laying the foundations for graspable user interfaces. In: *Proceedings of the SIGCHI Conference on Human Factors in Computing Systems CHI '95*, Denver, CC, USA, pp442–449, May 07–11, 1995.

Flannery, T. F., *The Future Eaters – an ecological history of the Australasian lands and people*. Reed New Holland, 1994.

Flood, J., *Archaeology of the Dreamtime – the story of prehistoric Australia and its people*. Revised edition, Gecko Books, 2010.

Freedman, D. H., Why scientific studies are so often wrong: The street light effect. *Discover Magazine*, July–August 2010.

Freedman, D. H., *Wrong - why experts keep failing us, and how to know when not to trust them*. Little, Brown and Company, 2010.

Fry, S. J., *Mythos - the Greek myths retold*. Penguin, 2017.

Fukasawa, N. (ed.), *Naoto Fukasawa*. Phaidon, 2007.

Fukumoto, M., and Toshiaki, S., Active click: Tactile feedback for touch panels. In: *Proceedings of the CHI 2001 Conference*, pp121–122, 2001.

Gammage, B., *The Biggest Estate on Earth – how Aborigines made Australia*. Allen & Unwin, 2011.

García-Bermejo, J. M. F., *The Other Side of Painting - Marcel Duchamp*. Ediciones Polígrafa, 1996.

Gehl, J. *Life Between Buildings – using public space*. 2011 Edition, Island Press, 1987.

Gershenfeld, N., *When Things Start to Think*. Hodder & Stoughton, 1999.

Gershenfeld, N., Krikorian, R., and Cohen, D. The Internet of Things. *Scientific American*, 291/4, pp76–81, October 2004.

Gibbons, A., A new kind of ancestor: Ardipithecus unveiled. *Science*, 326/5949, pp36–40, 2 October 2009.

Gibson, J. J., *The Perception of the Visual World*. Greenwood Press, 1950.

Gibson, J. J., and Crooks, L. E., A theoretical field-analysis of automobile-driving. *The American Journal of Psychology*, 51/3, pp543–471, July 1953.

Gibson, J. J., *The Senses Considered as Perceptual Systems*. Houghton Mifflin, 1966.

Gibson, J. J., The theory of affordances. In: Shaw, R., and Bransford, J. (eds.), *Perceiving, Acting, and Knowing – toward and ecological psychology*, pp67–82, Lawrence Erlbaum, 1977.

Gibson, J. J., *The Ecological Approach to Visual Perception*. Houghton Mifflin, 1979.

Gibson, K. R., and Ingold, T. (eds.), *Tools, Language and Cognition in Human Evolution*. Cambridge University Press, 1993.

Gibson, W., and Sterling, B. *The Difference Engine*. Bantam Books, 1991.

Gladwell, M., *Blink – the power of thinking without thinking*. Penguin, 2005.

Gladwell, M., Creation Myth - Xerox PARC, apple, and the truth about innovation. *The New Yorker*, 9 May 2011.

Gleick, J., *Chaos - making a new science*. Penguin Books, 1987.

Gleick, J., *The Information – a history, a theory, a flood*. Fourth Estate, 2011.

Godfrey-Smith, P., *Other Minds – the octopus and the evolution of intelligent life*. Harper Collins, 2016.

de Goede, J. (ed.), *Jean Tinguely. Alles Beweegt!* Kunsthal Rotterdam, Uitgeverij THOTH, 2007.

Goleman, D., *Emotional Intelligence - why it can matter more than IQ*. Bantam Books, 1995.

Goleman, D., *Focus – the hidden driver of excellence*. Bloomsbury, 2013.

Gombrich, E. H., *The Story of Art*. (First edition 1950) 16th edition, Phaidon, 1995.

Gombrich, E. H., *Art and Illusion – a study in the psychology of pictorial representation*. (First edition 1960) 6th edition, Phaidon, 2002.

Gombrich, E. H., *Little History of the World*. Yale, 2005, 2011.

Gould, S. J., *Bully for Brontosaurus*. Penguin, 1991.

Gould, S. J., *Eight Little Piggies – reflections in natural history*. Penguin, 1993.

Grandjean, E., *Fitting the Task to the Man: an ergonomic approach*. Taylor and Francis, 1980.

Grayling, A. C. *The Age of Genius – the seventeenth century and the birth of the modern mind*. Bloomsbury, 2016.

Gribbin, J. Alone in the milky way. *Scientific American*, 319/3, pp86–81, September 2018.

Gross, S., Bardzell, J., and Bardzell, S., Skeu the evolution: Skeuomorphs, style, and the material of tangible interactions. In: *Proceedings of the international conference on Tangible and Embedded Interaction (TEI)*, Munich, 16–19 February 2014.

Grudin, J., A Moving Target: the evolution of HCI. In: Jacko, J. A. (ed.), *Human-Computer Interaction Handbook - fundamentals, evolving technologies, and emerging applications*, ppxxvii–lxi. 3rd Edition, CRC Press, 2012.

Guallart, V. (ed.), *The Media House*. Actar, 2005.

Haas, R., Watson, J., Buonasera, T., Southon, J., Chen, J. C., Noe, S., Smith, K., Viviano Llave, C., Eerkens, J., and Parker, G., Female hunters of the early Americas. *Science Advances*, 6/45, 4 November 2020.

Hackel, M., Hot rocks, heat treatment at burrill lake and currarong, New South Wales. *Archaeology in Oceania*, 20/3, pp98–103, October 1985.

Hague, R., Campbell, I., and Dickens, P., Implications on design of rapid prototyping. In: *Proceedings of the Institution of Mechanical Engineers*, 217/Part C, pp25–30, 2003.

Hall, E. T., *The Hidden Dimension –man's use of space in public and private*. Anchor Books, 1966.

Hall, E. T., *The Dance of Life - the other dimension of time*. Anchor Books, 1983.

Hall, S., *This Means This, This Means That - a user's guide to semiotics*. 2nd edition, Laurence King Publishing, 2012.

Harari, Y. N., *Sapiens – a brief history of humankind*. Vintage, 2014.

Harari, Y. N., *Homo Deus – a brief history of tomorrow*. Vintage, 2016.

Harris, Y. C., and Bongers, A. J., Approaches to creating interactivated spaces, from intimate to inhabited interfaces. *Journal of Organised Sound*, Cambridge University Press, Special issue on Interactivity, 7/3, pp239–246, December 2002.

Hassett, L., van den Berg, M., Lindley, R. I., Crotty, M., McCluskey, A., Van Der Ploeg, H. P., Smith, S. T., Schurr, K., Killington, M., Bongers, B., and Howard, K., Effect of affordable technology on physical activity levels and mobility outcomes in rehabilitation: A protocol for the Activity and MObility UsiNg Technology (AMOUNT) rehabilitation trial. *British Medical Journal (BMJ) Open*, 6, June 2016.

Hassett, L., van den Berg, M., Lindley, R. I., Crotty, M., McCluskey, A., van der Ploeg, H. P., Smith, S. T., Schurr, K., Howard, K., Hackett, M. L., and Killington, M., Digitally enabled aged care and neurological rehabilitation to enhance outcomes with Activity and MObility UsiNg Technology (AMOUNT) in Australia: A randomised controlled trial, *PLoS Medicine*, 17/2, February 2020.

Haviland, J. B., Guugu Yimithirr cardinal directions. *Ethos*, 26/1, pp25–47. March 1998.

Heldring, L., Robinson J. A., and Vollmer, S., *Monks, Gents and Industrialists: The Long-Run Impact of the Dissolution of the English Monasteries*. National Bureau of Economic Research, working paper 21450, August 2015.

Hermann, T., Hunt, A., and Neuhoff, J. G. (eds.), *The Sonification Handbook*. Logos Verlag, 2011.

Heskett, J., *Toothpicks and Logos - design in everyday life*. Oxford University Press, (also by the same publisher, as *Design - a very short introduction*, in 2005), 2002.

Higham, T., Douka, K., Wood, R., Ramsey, C. B., Brock, F., Basell, L., Camps, M., Arrizabalaga, A., Baena, J., Barroso-Ruíz, C., and Bergman, C., The timing and spatiotemporal patterning of Neanderthal disappearance. *Nature*, 512, pp306–309, 21 August 2014.

Hiscock, P., O'Connor, S., Balme, J., and Maloney, T., World's earliest ground-edge axe production coincides with human colonisation of Australia. *Journal of Australian Archaeology*, 82/1, pp2–11, 2016.

Hix, D., and Hartson, H. R., *Developing User Interfaces Ensuring Usability Through Product & Process*. John Wiley & Sons, Inc., 1993.

Hochman, E. S., *Bauhaus - crucible of modernism*. Fromm International, 1997.

Hodge, R., and Kress, G. R., *Social Semiotics*. Polity Press, 1988.

Hoffman, D. L., Standish, C. D., García-Diez, M., Pettitt, P. B., Milton, J. A., Zilhão, J., Alcolea-González, J. J., Cantalejo-Duarte, P., Collado, H., De Balbín, R., and Lorblanchet, M., U-Th dating of carbonate crusts reveals Neanderthal origin of Iberian cave art. *Science*, 359, pp912–915, 23 February 2018.

Honing, H., *Iedereen is Muzikaal - wat we weten over het luisteren naar muziek.* Eerste druk 2009, herziene druk met nieuw nawoord (revised edition with new afterword), Nieuw Amsterdam, 2012.

Honing, H. Without it no music - Beat induction as a fundamental musical trait. *Annals of the New York Academy of Sciences*, 1252, pp85–91, 2012.

Honing, H., *Aap Slaat Maat - op zoek naar de oorsprong van muzikaliteit bij mens en dier.* Nieuw Amsterdam, 2018.

Hopper, A., Sentient computing. The Clifford Paterson lecture at the royal society 11 May 1999. *Philosophical Transactions of the Royal Society A: Mathematical, Physical and Engineering Sciences*, 358, pp2349–2358, 2000.

Hornecker, E., Beyond affordance: Tangibles' hybrid nature. In: *Proceedings of the Sixth International Conference on Tangible, Embedded and Embodied Interaction, TEI'12*, Kingston, ON, Canada, pp175–182, 19–22 February, 2012.

Horowitz, P., and Hill, W., *The Art of Electronics.* Cambridge University Press, 1980.

Huerta-Sánchez, E., Jin, X., Bianba, Z., Peter, B. M., Vinckenbosch, N., Liang, Y., Yi, X., He, M., Somel, M., Ni, P., and Wang, B., Altitude adaptation in Tibetans caused by introgression of Denisovan-like DNA. *Nature*, 512/7513, pp194–197, 2014.

Hughes, R. *The Fatal Shore – a history of the transportation of convicts to Australia, 1787–1868.* Vintage, 1986.

Hurlston, D. (ed.), *Ron Mueck.* NGV, 2010.

Hwang, F., Keates, S., Langdon, P., and Clarkson, P. J., Multiple haptic targets for motion-impaired users. In: *Proceedings of the CHI 2003 Conference on Human Factors in Computing Systems*, pp41–48, 2003.

Impett, J., A Meta-Trumpet(-er). In: *Proceedings of the International Computer Music Conference (ICMC)*, Århus, Denmark, pp147–150, 1994.

Impett, J., Projection and interactivity of musical structures in mirror-rite. *Journal of Organised Sound*, 1/3, pp203–211, 1996.

Impett, J., and Bongers, A. J., Hypermusic and the sighting of sound, a nomadic studio report. In: *Proceedings of the International Computer Music Conference*, Havana Cuba, 2001.

Ishii, H., and Ullmer, B., Tangible bits: Towards seamless interfaces between people, bits and atoms. In: *Proceedings of the ACM SIGCHI Conference on Human Factors in Computing Systems, CHI'97*, Atlanta, GA, USA, pp234–241, 22–27 March 1997.

Jacko, J. A. (ed.), *The Human-Computer Interaction Handbook - fundamentals, evolving technologies, and emerging applications.* 3rd edition, CRC Press, 2012.

Jakobson, R., Linguistics and poetics. In: Sebeok, T. A. (ed.), *Style in Language*, pp350–377, Wiley, 1960.

Jones, J. S., *Darwin's Island – the Galapagos in the garden of England.* Little, Brown, 2009.

Jormakka, K., *Flying Dutchmen, Motion in Architecture.* Birkhäuser, 2002.

Jennings, H., *Pandæmonium –1660–1886: the coming of the machine as seen by contemporary observers.* Pan Books, 1985.

Jewitt, C. (ed.), *The Routledge Handbook of Multimodal Analysis*. Routledge, 2009.

Kahneman, D., *Thinking, Fast and Slow*. Penguin, 2011.

Kalawsky, R. S., *The Science of Virtual Reality and Virtual Environments*. Addison Wesley, 1993.

Kay, A., Computer software. *Scientific American*, 251/3, pp53–59, September 1984.

Kay, A., User interface, a personal view. In: Laurel, B. (ed.), *The Art of Human-Computer Interface Design*, pp191–207, Addison-Wesley, 1990.

Käufer, S., and Chemero, A. *Phenomenology – an introduction*. Polity Books, 2015.

Kayzer, W. (ed.), *Een Schitterend Ongeluk (a glorious accident)*. VPRO, 1993.

Kelly, L., *The Memory Code - the traditional Aboriginal memory technique that unlocks the secrets of Stonehenge, Easter Island and ancient monuments the world over*. Allen & Unwin, 2016.

Kelly, L., *Memory Craft - improve your memory using the most powerful methods from around the world*. Allen & Unwin, 2019.

Klenerman, L., *Human Anatomy - a very short introduction*. Oxford University Press, 2015.

Kohn, M., and Mithen, S. J., Handaxes: Products of sexual selection? *Antiquity*, 73, pp518–526, 1999.

Kolko, J., *Thoughts on Interaction Design*. 2nd edition, Morgan-Kaufmann, 2011.

Krause, J., Fu, Q., Good, J. M., Viola, B., Shunkov, M. V., Derevianko, A. P., and Pääbo, S., The complete mitochondrial DNA genome of an unknown hominin from southern Siberia. *Nature*, 464, pp894–897, 8 April 2010.

Krefeld, V., The hand in the web – An interview with Michel Waisvisz. *Computer Music Journal*, 14/2, pp28–33, Summer 1990.

Kress, G. R., Against arbitrariness – The social production of the sign as a foundational issue in critical discourse analysis. *Discourse & Society*, 4/2, pp169–191, 1993.

Kress, G. R., and van Leeuwen, T. J., *Multimodal Discourse, the modes and media of contemporary communication*. Oxford University Press, 2001.

Kress, G. R., and van Leeuwen, T. J., *Reading Images – the grammar of visual design*. 2nd edition, Routledge, 2006.

Kress, G. R., *Multimodality – a social semiotic approach to contemporary communication*. Routledge, 2010.

Kurzweil, R., *The Age of Spiritual Machines - when computers exceed human intelligence*. Penguin, 1999.

Lakoff, G., and Johnson, M., *Metaphors We Live By*. The University of Chicago Press, 1980.

Landes, D. S., *The Unbound Prometheus – technological change and industrial development in Western Europe from 1750 to the present*. Cambridge University Press, 1969.

Lane, M. (ed.) *Structuralism - a reader*. Jonathan Cape Ltd., 1970.

Langdon, P. Keates, S., Clarkson, J., and Robinson, P., Using haptic feedback to enhance computer interaction for motion-impaired users. In: *Proceedings of the International Conference on Disability*, Virtual Reality and Associated Technologies, Sardinia, Italy, 2000.

Lasserre, B., *Words That Go "Ping" - the ridiculously wonderful world of onomatopoeia*. Allen & Unwin, 2018.

Laurel, B. (ed.,) *The Art of Human-Computer Interface Design*. Addison-Wesley, 1990.

Laurel, B. *Computers as Theatre*, Addison-Wesley, 1991.

Laurel, B. (ed.), *Design Research - methods and perspectives*. MIT Press, 2003.

Lawlor, R. *Voices of the First Day - awakening in the Aboriginal dreamtime*. Inner Traditions International, 1991.

LeDoux, J. E., *The Emotional Brain – the mysterious underpinnings of emotional life*. Phoenix, 1996.

van Leeuwen, T. J., *Speech, Music, Sound*. Macmillan Press, 1999.

Licklider, J. C. R., Man-computer symbiosis. In: *IRE Transactions on Human Factors in Electronics*, pp4–11, March 1960.

Lie, S., *Assisting Product Designers with Balancing Strength and Surface Texture of Handheld Products Made from 3D Printed Polymers*. Ph.D. thesis, UTS, July 2020.

Lim, K. Y., and Long, J., *The MUSE Method for Usability Engineering*. Cambridge University Press, 1994.

Lima, M., *The Book of Trees – visualising branches of knowledge*. Princeton Architectural Press, 2014.

Lovelock, J. (with Appleyard, B.), *Novacene - the coming age of hyperintelligence*. Penguin, 2019.

Löwgren, J., Interaction design - Brief intro. In: Soegaard, M., and Dam, R. F. (Eds), *The Encyclopedia of Human-Computer Interaction*. 2nd edition, Chapter 1, Interaction Design Foundation, 2014.

Luyben, L., and Posthouwer, I., *Small Talk Survival – praktische gids voor de gesprekken tussendoor*. Ambo|Anthos, 2019.

Lynn, G., *Muscle NSA* interview with Kas Oosterhuis by Greg Lynn. Archaeology of the Digital, 5, Canadian Centre for Architecture, 2014.

Lynn, G., *H₂O Expo* interview with Lars Spuybroek by Greg Lynn. Archaeology of the Digital, 10, Canadian Centre for Architecture, 2015.

Lyons, J., *Semantics*. Cambridge University Press, 1977.

Mandrelli, D. O. (ed.), Less aesthetics more ethics. In: *Catalogue of the Biennale di Venezia, 7th International Architecture Exhibition*, Venice, 2000.

Marchand, P., *Marshall McLuhan, the Medium and the Messenger*. MIT Press (1998), 1989.

Markillie, P., Special report: Manufacturing and innovation, *The Economist*, 403/8781, 21 April 2012.

Martinelli, D., *A Critical Companion to Zoosemiotics - people, paths, ideas*. Springer, 2010.

Marzke, M. W., Precision grips, hand morphology, and tools. *American Journal of Physical Anthropology*, 102, pp91–110, 1997.

McCullough, M., *Abstracting Craft, The Practised Digital Hand*. MIT Press, 1996.

McGrady, P., *The Machine that Made Us*. Wavelength Films Ltd., 2008.

McLuhan, E., Kuhns S., and Cohen, M. (eds.), *Marshall McLuhan: The Book of Probes*. Ginko Press, 2003.

McLuhan, H. M., *The Mechanical Bride – folklore of industrial man*. Duckworth Overlook, 1951.

McLuhan, H. M., Culture without literacy. *Explorations in Communication*, 1, 1953 (Reprinted in: McLuhan, E., and Gordon, W. T., Marshall McLuhan Unbound, 6. Ginko Press, 2005).

McLuhan, H. M., The effect of the printed book on language in the 16th century. *Explorations in Communication*, 7, 1957 (Reprinted in: McLuhan, E., and Gordon, W. T., Marshall McLuhan Unbound, 2. Ginko Press, 2005).

McLuhan, H. M., The medium is the message, *Forum*, pp19–24, Spring 1960 (Reprinted in: McLuhan, E., and Gordon, W. T., Marshall McLuhan Unbound, 17. Ginko Press, 2005).

McLuhan, H. M., *The Gutenberg Galaxy – the making of typographic man*. University of Toronto Press, 1962.

McLuhan, H. M., *Understanding Media – the extensions of man*. Routledge (Routledge Classics edition 2001), 1964.

McLuhan, H. M., and Fiore, Q., *The Medium is the Massage - an inventory of effects*. Bantam Books, 1967 (Reprinted with a different cover by Ginko Press, 2001).

McLuhan, H. M., *Counterblast*. Harcourt, Brace & World Inc., 1969.

Mery-Keitel, A. S., *Human-Computer Interaction in Museums as Public Spaces. A research of the impact of interactive technologies on visitors' experience*, Ph.D. thesis, University of Technology Sydney, 2013.

Millard, K., *Milgram Was Wrong: We Don't Obey Authority, But We Do Love Drama*. The Conversation, 4 February 2015.

Miranda, E. R., and Wanderley, M. M., *New Digital Musical Instruments: Control and Interactions Beyond the Keyboard*. A-R Editions, 2006.

Mithen, S. J., *The Prehistory of the Mind - a search for the origins of art, religion and science*. Phoenix, 1996.

Mithen, S. J., *The Singing Neanderthals – the origins of music, language, mind and body*. Phoenix, 2005.

Mlodinow, L. *Subliminal – the new unconscious and what it teaches us*. Penguin, 2012.

Moggridge, W. G., *Designing Interactions*. MIT Press, 2007.

Moore, G. E., Cramming more components onto integrated circuits. *Electronics*, 38/8, pp114–117, April 1965.

Morris, C., Foundations of the theory of signs. Original in 1938, reprinted in: Morris, C. (ed.), *Writings on the General Theory of Signs*. Part I, pp17–74, Mouton, 1971.

Morris, C., Signs, Language, and Behavior. Original in 1946, reprinted in: Morris, C. (ed.), *Writings on the General Theory of Signs*. Part II, pp75–400. Mouton, 1971.

Mueller, F., Edge, D., Vetere, F., Gibbs, M. R., Agamanolis, S., Bongers, A. J., and Sheridan, J. G., Designing sports: A framework for exertion games. In: *Proceedings of the SIGCHI Conference on Human Factors in Computing Systems (CHI '11)*, pp2651–2660, ACM, 2011.

Mueller, F., Toprak, C., Graether, E., Walmink, W., Bongers, A. J., and van den Hoven, E. A. W. H., Hanging off a bar. In: *CHI '12 Extended Abstracts on Human Factors in Computing Systems (CHI EA '12)*, pp1055–1058, ACM, 2012.

Mulder, A., *Over Mediatheorie - taal, beeld, geluid, gedrag*. NAI Publishers, 2004.

Mulder, J., Functions of amplified music: A theoretical approach. In: *Proceedings of 20th International Congress on Acoustics*, Sydney, 2010.

Mulder, J., *Making Things Louder - amplified music and multimodality*. Ph.D. Thesis, University of Technology Sydney, 2013.

Mumford, L., *Technics and Civilisation*. Routledge, 1934.

Mumford, L., *The City in History – its origins, its transformations, and its prospects*. Penguin, 1961.

Mumford, L., *The Myth of the Machine – technics and human development*. Secker and Warburg, 1967.

Murell, K. F. H., *Ergonomics: Man and His Working Environment*. Chapman and Hall Ltd., 1965.

Myers, F. R., *Pintupi Country, Pintupi Self: sentiment, place, and politics among western desert Aborigines*. University of California Press, 1986.

Neale, M., and Kelly, L., *Songlines - the power and promise*. Thames & Hudson, 2020.

Nelson, T. H., Complex information processing: A file structure for the complex, the changing, and the indeterminate. In: Wardrip-Fruin, N., and Montfort, N. (eds.), *The New Media Reader*, pp134–145, MIT Press, 2003. Originally in: *Proceedings of the 20th ACM National Conference*, pp84–100, August 1965.

Nelson, T. H., The right way to think about software design. In: Laurel, B. (ed.), *The Art of Human-Computer Interface Design*, pp235–243, Addison-Wesley, 1990.

Neisser, U. G., *Cognition and Reality – principles and implications of cognitive psychology*. W. H. Freeman, 1976.

Nicholson, I., Milgram goes to hollywood. *American Psychological Society PsycCRITIQUES*, 61/30, 25 July 2016.

Nielsen, J., *Usability Engineering*. Academic Press Professional, 1993.

New, T., The niche. In: Attiwill, P., and Wilson, B. (eds.), *Ecology, an Australian Perspective*. Chapter 12, pp184–194. Oxford University Press, 2003.

Norman, D. A., Cognitive engineering. In: Norman, D. A., and Draper, S. W. (eds.), *User Centered System Design - new perspectives on Human-Computer Interaction*, Chapter 3. Lawrence Erlbaum, 1986.

Norman, D. A., *The Psychology of Everyday Things*. Basic Books, 1988.

Norman, D. A., *The Invisible Computer: why good products can fail, the personal computer is so complex, and information appliances are the solution*. MIT Press, 1998.

Norman, D. A., *Emotional Design – why we love (or hate) everyday things*. Basic Books, 2004.

Norman, D. A., *The Design of Everyday Things*. 2nd (revised & expanded) edition, Basic Books, 2013.

Nørretranders, T., *The User Illusion - cutting consciousness down to size*. Translated from the Danish *Mærk Verden - En beretning om bevidsthed* (1991) by Jonathan Sydenham, Penguin, 1998.

Norris, S., *Analyzing Multimodal Interaction - a methodological framework*. Routledge, 2004.

O'Connell, M., *To Be a Machine - adventures among cyborgs, utopians, hackers, and the futurists solving the modest problem of death*. Granta Books, 2016.

Ogden, C. K., and Richards, I. A., *The Meaning of Meaning – a study of the influence on language upon thought and of the science of symbolism* (first published in 1923). 8th edition, Routledge & Kegan Paul Ltd., 1946.

Oosterhuis, K., *Architecture Goes Wild*. 010 Publishers, 2002.

Oosterhuis, K., *Hyperbodies – towards an E-motive architecture*. Birkhäuser, 2003.

Oosterhuis, K., *Towards a New Kind of Building- a designer's guide for nonstandard architecture*. NAI Publishers, 2011.

Oosterhuis, K., Bier, H., Biloria, N., Kievid, C., Slootweg, O., and Xia, X. (eds.), *Hyperbody - first decade of interactive architecture*. Jap Sam Books, 2012.

Organ, C., Nunn, C. L., Machanda, Z., and Wrangham, R. W., Phylogenetic rate shifts in feeding time during the evolution of *Homo*. *Proceedings of the National Academy of Sciences*, 108/35, pp14555–14559, 30 August 2011.

O'Sullivan, D., and Igoe, T., *Physical Computing – sensing and controlling the physical world with computers*. Thomson Course Technology PTR, 2004

Paradiso, J., New ways to play: Electronic music interfaces. *IEEE Spectrum*, 34/12, cover article and pp18–30, December 1997.

Pascoe, B., *Dark Emu – Aboriginal Australia and the birth of agriculture*. 2nd edition, Magabala Books, 2018.

Papanek, V., *Design for the Real World - made to measure*. Thames & Hudson, 1971.

Perrella, S. (ed.), Hypersurface architecture II. In: *Architectural Design*, 69/9–10, John Wiley &Sons, 1999.

Perry, G., *Behind the Shock Machine - the untold story of the notorious Milgram psychology experiments*. The New Press, 2013.

Perry, G., The shocking truth of the notorious Milgram obedience experiments. *Discover Magazine*, 3, Octber 2013.

Petrilli, S., and Ponzio, A., *Thomas Sebeok and the Signs of Life*. Icon Books, 2001.

von Petzinger, G., *The First Signs - unlocking the mysteries of the world's oldest symbols*. Atria Paperback, 2016.

Pheasant, S., *Bodyspace - anthropometry, ergonomics and the design of work*. 2nd edition, Taylor & Francis, 1996.

Pickrell, M., Bongers, A.J., and van den Hoven, E. A. W. H., Understanding persuasion and motivation in interactive stroke rehabilitation. In: *Proceedings of of the 10th International Conference on Persuasive Technology*, Chicago, pp15–26, Springer LNCS, 2015.

Pickrell, M., Bongers, A. J., and van den Hoven, E. A. W. H., Understanding changes in motivation of stroke patients undergoing rehabilitation in hospital. In: *Proceedings of the 11th International Conference on Persuasive Technology*, Salzburg, pp251–262, Springer LNCS, 2016.

Pickrell, M., van den Hoven, E. A. W. H., and Bongers, A. J., Exploring in-hospital rehabilitation exercises for stroke patients: Informing interaction design. In: *Proceedings of the 29th Australian Conference on Human Computer Interaction*, pp228–237, ACM Press, 2017.

Pickrell, M., *The Design of Interactive Technology for Stroke Patient Rehabilitation*. Ph.D. thesis, University of Technology Sydney, 2020.

Pinker, S. *The Language Instinct – the new science of language and mind*. Penguin, 1994.

Pinker, S., *How the Mind Works*. Norton, 1997.

Platt,C., *Make: Electronics - learning by discovery*. O'Reilly Media Inc., 2009.

Pocock, G., and Richards, C., *Human Physiology - the basics of medicine*. 2nd edition, Oxford University Press, 2004.

Postman, N., *Technopoly – the surrender of culture to technology*. Vintage, 1992.

Potter, P., Mister self destruct - Shredded art. In: *Banksy - you are an acceptable level of threat and if you were not you would know about it*. 10th edition. p245, Carpet Bombing Culture, 2019.

Rappolt, M., *Greg Lynn FORM*. Rizzoli International Publications, 2008.

Raskin, J., *The Humane Interface - new directions for designing interactive systems*. ACM Press, 2000.

Rasmussen, J., *Information Processing and Human-Machine Interaction - an approach to cognitive engineering*. North-Holland, 1986.

Read, D., The ties that bind. *Nature*, 388, pp517–518, 7 August 1997.

Rheingold, H., *Virtual Reality*. Mandarin Paperbacks, 1991.

Robinson, A., *The Story of Writing - alphabets, hieroglyphs & pictograms*. Thames & Hudson, (first edition 1995), new edition 2007.

Rolls, E. C., No fixed address – Response. *Quarterly Essay*, 25, pp61–65, 2007.

Rosengren, K. E., *Communication: an introduction*. Sage, 2000.

Rosenthal, N. (ed.), *Sensation: Young British Artists from the Saatchi Collection*. Royal Academy of Arts, London, Thames & Hudson, 1997.

Ross, S., *The Science of Sherlock Holmes*. Michael O'Mara Books Ltd., 2020.

de Ruyter, B. (ed.), *365 Days' Ambient Intelligence Research in HomeLab*. Philips Research, 2003.

Sacks, O. W., *The Man Who Mistook his Wife for his Hat*, Picador, 1985.

Sacks, O. W., *Musicophilia - tales of music and the brain*. Picador, 2007.

Salvendy, G. (ed.), *Handbook of Human Factors and Ergonomics*. 4th edition, Wiley & Sons, 2012.

Schafer, R. M., *The Soundscape – our sonic environment and the tuning of the world*. Destiny Books, 1977.

Schneiderman, B., *Designing the User Interface: Strategies for Effective Human-Computer Interaction*. Addison-Wesley, 1986.

Schomaker, L., S. Münch, and K. Hartung (eds.), *A Taxonomy of Multimodal Interaction in the Human Information Processing System*. Report of the ESPRIT Project 8579. MIAMI, 1995.

Scolari, C., Towards a semio-cognitive theory of HCI. In: *Extended Abstracts of the ACM CHI 2001 Conference*, pp85–86, ACM Press, 2001.

Scolari, C., The sense of the interface: Applying semiotics to HCI research, *Semiotica* 177–1/4, pp1–27, 2009.

Sebeok, T. A., Hayes, A. S., and Bateson, M. C. (eds.), *Approaches to Semiotics - transactions of the Indiana University conference on paralinguistics and kinesics*. Mouton, 1964.

Sebeok, T. A., Animal communication - A communication network model for languages is applied to signalling behaviour in animals. *Science*, 147, pp1006–1014, 1965.

Sebeok, T. A. *Perspectives in Zoosemiotics*. Mouton, 1972.

Sebeok, T. A., and Danesi, M., *The Forms of Meaning: modeling systems theory and semiotic analysis*. Mouton de Gruyter, 2000.

Shaer, O., and Hornecker, E., Tangible User Interfaces – Past, present, and future directions. *Foundations and Trends in Human-Computer Interaction*, 3/1–2, pp1–137, 2009.

Shannon, C. E., and Weaver, W., *The Mathematical Theory of Communication*. University of Illinois Press, 1949.

Shariatmadari, D., Why it's time to stop worrying about the decline of the English language. *The Guardian*, 15, August 2019.

Sharp, H., Rogers, Y., and Preece, J., *Interaction Design - beyond human-computer interaction*. 1st edition 2002, 2nd edition, 2007, 5th edition, John Wiley & Sons, 2019.

Shea, J. J., Lithic Modes A – I, A new framework for describing global-scale variation in stone tool technology illustrated with evidence from the east mediterranean levant. *Journal of Archaeological Method and Theory*, 20, pp151–186. 2013.

Sheldrake, R., *Science and Spiritual Practices - reconnecting through direct experience*. Coronet, 2017.

Sheldrake, M., *Entangled Life - how fungi make our worlds, change our minds, and shape our futures*. Penguin, 2020.

Sibling, E., *The Cello Suites - J. S Bach, Pablo Casals, and the search for a baroque masterpiece*. Allen & Unwin, 2009.

Simard, S. W., Perry, D. A., Jones, D. M., Myrold D. D., Durall D. M., and Molina, R., Net transfer of carbon between ectomycorrhizal tree species in the field. *Nature*, 388, pp579–582, 7 August 1997.

van de Sluis, R., Bongers, A. J., Kohar, H., Jansen, J., Pauws, S. C., Eggen, J. H., and Eggenhuisen, H., *WWICE User Interface Concepts*. Philips NatLab Technical Note, 1997.

van de Sluis, R., Eggen, J. H., Kohar, H., Jansen, J., User interface for an in-home environment. In: *Proceedings of the Interact Conference*, Tokyo, 2001.

Smit, T., *Eden*. Transworld Publishers, 2001.

Smith, D. C., Irby, C., Kimhall, R., and Verplank, W. L., Designing the Star user interface. *Byte*, pp242–282, April 1982.

Smith, M. (ed.), *Stelarc - the monograph*. MIT Press, 2005.

Snow, C. P., *The Two Cultures and A Second Look - an expanded version of the two cultures and the scientific revolution*. Cambridge University Press, 1964.

de Souza, C. S., Barbosa, S. B. J., and da Silva, S. R. P., Semiotic engineering principles for evaluating end-user programming environments. *Interacting with Computers*, 13, pp467–495, 2001.

de Souza, C. S., *The Semiotic Engineering of Human-Computer Interaction*. The MIT Press, 2005.

de Souza, C. S., and Leitão, C. F., *Semiotic Engineering Methods for Scientific Research in HCI*. Synthesis Lectures on Human-Centered Informatics #2, Morgan & Claypool, 2009.

Spiering, H., *Sapiens komt uit een Afrikaanse wolk*. NRC, ppW6–7, 6 June 2020.

Spuybroek, L., *NOX: Machining Architecture*. Thames & Hudson, 2004.

Standley, P.-M., Bidwell, N. J., George Senior, T., Steffensen, V., and Gothe, J. Connecting communities and the environment through media: Doing, saying and seeing along traditional Indigenous knowledge revival pathways. *3C Media: Journal of Community, Citizens and Third Sector Media and Communication*, 5, pp9–27, October 2009.

Staal, G. (ed.), *DieTwee- new Dutch graphic design*. BIS Publishers, 2002.

Steffensen, V., *Fire Country - how Indigenous fire management could help save Australia*. Hardie Grant Travel, 2020.

Sterling, B., When our environments become really smart. In: Denning, P. J. (ed.), *The Invisible Future, the seamless integration of technology into everyday life*. McGraw-Hill, 2002.

Sterling, B., *Shaping Things*. MIT Press, 2005.

Sterling, B., *The Epic Struggle of the Internet of Things*. Strelka Press, 2014.

Stewart, M., Clark-Wilson, R., Breeze, P. S., Janulis, K., Candy, I., Armitage, S. J., Ryves, D. B., Louys, J., Duval, M., Price, G. J., Cuthbertson, P., Bernal, M. A., Drake, N. A., Alsharekh, A. M., Zahrani, B., Al-Omari, A., Roberts, P., Groucutt, H. S., and Petraglia, M. D., Human footprints provide snapshot of last interglacial ecology in the Arabian interior. *Science Advances*, 6/38, 18 September 2020.

Stigler, S. M., Stigler's law of eponymy. In: *Science and Social Structure- a Festschrift for Robert K. Merton*. Transactions of the New York Academy of Social Sciences, Series II, 39/1 Series II, pp146–157, 1980.

Sutherland, I. E., Sketchpad - A man-machine graphical communication system. In: Wardrip-Fruin, N., and Montfort, N. (eds.), *The New Media Reader*. MIT Press, pp111–126, 2003. Originally in *American Federation of Information Processing Systems (AFIPS) Conference Proceedings* 23, pp329–346, 1963.

Sutikna, T., Tocheri, M. W., Morwood, M. J., Saptomo, E. W., Awe, R. D., Wasisto, S., Westaway, K. E., Aubert, M., Li, B., Zhao, J. X., and Storey, M., Revised stratigraphy and chronology for *Homo floresiensis* at Liang Bua in Indonesia. *Nature*, 532, pp366–369, 16 April 2016.

Tanaka, A., and Bongers, A. J., Global string: A musical instrument for hybrid space. In: Fleischmann, M., Strauss, W., (eds), *Proceedings of the Cast01 // Living in Mixed Realities*, pp177–181, MARS Exploratory Media Lab, FhG - Institut Medienkommunikation, 2001.

Tocheri, M. W., Unknown human species found in Asia. *Nature*, 568, pp176–178, 2019.

Torre, G., Andersen, K., and Baldé, F., The Hands: The making of a digital musical instrument. *Computer Music Journal*, 40/2, pp22–34, Summer 2016.

Trentmann, F., *Empire of Things – how we became a world of consumers, from the fifteenth century to the twenty-first*. Allen Lane/Penguin, 2016.

Turney, J., *I, Superorgamism – learning to love your inner ecosystem*. Icon Books, 2015.

Twaites, D., Escape Velocity: between the possible and the actual. In: Bullock, N., and French B. (eds.), *Shaun Gladwell: Pacific Undertow*. Catalogue for the Exhibition at the Museum of Contemporary Art (MCA) Sydney, 19 July–7 October 2019.

Valtolina, S., Baricelli, B. R., and Dittrich, Y., Participatory knowledge-management design: A semiotic approach. *Journal of Visual Languages and Computing*, 23, pp103–115, 2012.

Verplank, W. L., Interaction design sketchbook - Frameworks for designing inter-active products and systems. www.billverplank.com, 1 December 2009.

de Waal, F. B. M., *Are We Smart Enough to Know How Smart Animals Are?* Granta Publications, 2016.

de Waal, F. B. M., *Mama's Last Hug - animal emotions and what they teach us about ourselves.* Granta Publications, 2019.

Waisvisz, M. The Hands, a set of remote MIDI-controllers. In: *Proceedings of the International Computer Music Conference*, pp313–318, 1985.

Wanderley, M. M., and Battier, M. (eds.), *Trends in Gestural Control of Music.* IRCAM, 2000.

Ward, L., *Love Lace.* Exhibition Catalogue of the Finalists for the International Lace Award, Powerhouse Museum, 2011.

Wardrip-Fruin, N., and Montfort, N. (eds.), *The New Media Reader.* MIT Press, 2003.

Waters, C. N., Zalasiewicz, J., Summerhayes, C., Barnosky, A. D., Poirier, C., Gałuszka, A., Cearreta, A., Edgeworth, M., Ellis, E. C., Ellis, M., and Jeandel, C., The anthropocene is functionally and stratigraphically distinct from the Holocene. *Science*, 351/6269, p137, 8 January 2016.

Weiser, M. The computer for the 21st century. *Scientific American*, 265/3, pp94–104, September 1991.

Whitehead, K., *New Dutch Swing - an in-depth examination of Amsterdam's vital and distinctive jazz scene.* Billboard, 1998.

Whitford, F., *Bauhaus.* Thames & Hudson World of Art Series, 1984.

Wickens, C. D., and Kramer, A., Engineering psychology. *Annual Review of Psychology*, 36, pp307–348, 1985.

Wickens, C. D., *Engineering Psychology and Human Performance.* 2nd edition, HarperCollins Publishers Inc., 1992.

Wickens, C. D., Holland, J. G., Banbury, S., and Parasumaran, R., *Engineering Psychology and Human Performance.* 4th edition, Routledge, 2013.

Wiggins, B. E., and Bowers, G. B., Memes as genre: A structurational analysis of the memescape. *New Media & Society*, 17/11, pp1886–1906, 2015.

Wigley, M, *Constant's New Babylon - the hyper-architecture of desire.* Witte de With, 1998.

Wilson, F. R., *The Hand – how it shapes the brain, language, and human culture.* Vintage, 1998.

Winawer, J., Witthoft, N., Frank, M. C., Wu, L., Wade, A. R., and Boroditsky, L., Russian blues reveal effects of language on color discrimination. *Proceedings of the National Academy of Sciences (PNAS)*, 104/19, pp7780–7785, 8 May 2007.

Winkler, I., Háden, G. P., Ladinig, O., Sziller, I., and Honing, H., Newborn infants detect the beat in music. *PNAS*, 107/7, pp2468–2471, 2009.

Wong, K., Last hominin standing. *Scientific American*, 319/3, pp56–61, September 2018.

Wood, D. J., *How Children Think and Learn - the social contexts of cognitive development.* 2nd edition, Blackwell, 1998.

Wrangham, R. W., *Catching Fire - how cooking made us human.* Profile Books, 2009.

Wray, A., Protolanguage as a holistic system for social interaction. *Language & Communication Journal*, 18/1, pp47–67, 1998.

Yunkaporta, T., *Sand Talk - how indigenous thinking can save the world*. The Text Publishing Company, 2019.

Zachar, I., The feasibility of segmentation of protolanguage. *Interaction Studies Journal*, 12/1, pp1–35, 2011.

Zeberg, H., and Pääbo, S., The major genetic risk factor for severe COVID-19 is inherited from Neanderthals. *Nature*, 587, pp610–612, 26 November 2020.

Zellner, P., *Hybrid Space - new forms in digital architecture*, Thames & Hudson, 1999.

Index

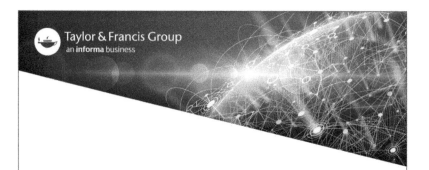